T0228291

The Myth of the Family Farm

Also of Interest

†*Science, Agriculture, and the Politics of Research*, Lawrence Busch and William B. Lacy

†*Politics in Rural America: People, Parties, and Policy*, Frank M. Bryan

†*Poverty in Rural America: A Case Study*, Janet M. Fitchen

An Introduction to the Sociology of Rural Development, Norman Long

Rural Society Research Issues for the 1980s, edited by Don A. Dillman and Daryl S. Hobbs

†*The Family in Rural Society*, edited by Raymond T. Coward and William M. Smith

†Available in hardcover and paperback.

A Westview Special Study

The Myth of the Family Farm:
Agribusiness Dominance of U.S. Agriculture
Ingolf Vogeler

The ideal of the family farm has been used to justify a myriad of federal farm legislation. Land grants, the distribution of irrigation water, land-grant college research and services, farm programs, and tax laws all have been affected. Yet, asserts the author, federal legislation and practices have had an institutional bias toward large-scale farms and agribusiness and have hastened the demise of family farms.

Dr. Vogeler examines the struggle between land interests in the private and public sectors and finds that the myth of the family farm has been used to obscure the dominance of agribusiness and that the corporate penetration of agriculture has in turn contributed to the plight of migrant workers, the decline of small towns, and the economic difficulties of independent farmers. Dr. Vogeler also identifies the major shortcomings of agribusiness and federal land-related laws and programs; examines the regional impact of agribusiness and federal farm programs on rural areas; and considers the role of racial minorities and women in the development of agrarian capitalism. In conclusion, he offers a structural analysis that provides the means for progressive social change and states that the achievement of economic equality in rural America and the dismantling of the corporate control of agriculture can be realized through farmer-labor alliances.

Ingolf Vogeler, associate professor of geography at the University of Wisconsin–Eau Claire, has conducted field research and published extensively for ten years on the underdevelopment and cultural landscapes of U.S. rural areas. He coedited *Dialectics of Third World Development* and is series editor for Westview's Geographies of the United States series.

The Myth of the Family Farm:
Agribusiness Dominance
of U.S. Agriculture

Ingolf Vogeler

Routledge
Taylor & Francis Group

LONDON AND NEW YORK

First published 1981 by Westview Press, Inc.

Published 2019 by Routledge
52 Vanderbilt Avenue, New York, NY 10017
2 Park Square, Milton Park, Abingdon, Oxon OX14 4RN

Routledge is an imprint of the Taylor & Francis Group, an informa business

Copyright © 1981 by Ingolf Vogeler

All rights reserved. No part of this book may be reprinted or reproduced or utilised in any form or by any electronic, mechanical, or other means, now known or hereafter invented, including photocopying and recording, or in any information storage or retrieval system, without permission in writing from the publishers.

Notice:
Product or corporate names may be trademarks or registered trademarks, and are used only for identification and explanation without intent to infringe.

Library of Congress Cataloging in Publication Data
Vogeler, Ingolf.
    The myth of the family farm.
    (A Westview special study)
    Includes index.
    1. Agriculture—Economic aspects—United States.   2. Agriculture and state—United States. 3. Family farms—United States. 4. Agricultural industries—United States. I. Title.
HD1761.V63 1981   338.1'6   80-21091

ISBN 13: 978-0-367-29419-9 (hbk)
ISBN 13: 978-0-367-30965-7 (pbk)

To a new Populism in
our lifetime

Contents

Tables

Figures

Acknowledgments

I had at first wanted to write this book for the Resource Paper Series of the Association of American Geographers. But because my proposal offered an alternative explanation of the status quo in U.S. agriculture, the reviewers felt that such divergent material would be inappropriate for geography undergraduates! Fortunately, Lynne Rienner, associate publisher for Westview Press, disagreed. Knowledge and ultimately truth are derived from intellectual debate rather than from monolithic viewpoints.

Philip Raup, land economist at the University of Minnesota, first taught and inspired me to search out alternative perspectives to the accepted virtues and inevitability of agribusiness. His curiosity about ideas rather than facts (the reverse unfortunately characterizes too many academics) was the fountainhead of my own self-education on rural matters. I thank him for being a fine teacher.

Several people assisted with this book and I thank them for their help: Chris LaLiberte for being able to read my handwriting and typing the first draft of the manuscript, Diane Frey for her excellent typing of the final copy of the book, and David Roggenbauer for making many of the maps and graphs. Sharon Knopp's usual excellent editorial assistance contributed substantively to the quality of this book, and Tony de Souza's careful reading and suggestions are greatly appreciated. Both Tony and Brady Foust of the Department of Geography at the University of Wisconsin–Eau Claire provided the supportive intellectual environment that encouraged me to write this book. Finally, the time spent with Teilen Vogeler-Knopp made a valuable contribution to my well-being during the writing of this book.

I. V.

Part 1

Introduction:
The Myth of the Family Farm

Figure 1.1. Agrarian democracy: the ideal and the real. (Top) The ideal of agrarian democracy was based on the family farm, but by the late nineteenth century, "independent" farm families were already dependent on agribusiness—in this photo, on the Chicago-based McCormick Reaper Company. State Historical Society of Wisconsin and USDA photo. (Bottom) The unfulfilled dreams of agrarian democracy: nineteenth-century land-granting laws prevented millions of immigrants, such as these in New York City about 1906, from settling the land but did provide for cheap labor in the booming cities. Library of Congress and USDA photo.

1
The Myth of the Family Farm

U.S. agriculture has been fundamentally transformed in the last fifty years. In 1920, when the farm population was first enumerated separately, about one American in three lived on farms. By 1977, about one person out of twenty-eight, or 3.6 percent of the nation's 216 million people, had a farm residence. This transformation has meant a massive shift in the farm population to urban centers. From 1920 to 1977 the net outmigration of the farm population totaled 48.7 million! From 1960 to 1970 alone the farm population declined from 15.6 million to 8 million, a 50 percent drop.[1] With a declining farm population and with the amount of farmland remaining constant, the average size of farms has increased (see Figure 1.2) from 150 acres in 1920 to 440 acres in 1979. Yields per acre and overall farm production also increased.

The conventional explanation for these agricultural changes is that inefficient farm operators have been eliminated by market forces while the remaining farmers are sufficiently organized to have survived. This viewpoint further asserts that U.S. agriculture is dominated by family farmers who run independent family-operated businesses. With the elimination of inefficient producers and the expansion of production, food consumers have benefited from lower food costs. In other words, capitalism and economic competition work!

This explanation of U.S. agriculture is widely accepted because it rests on the myth that capitalism is the best economic system for all concerned. Capitalism has indeed worked, but not for the benefit of the remaining family and tenant farmers, agricultural workers, food consumers, and small towns dependent on agriculture. Furthermore, the notion of a private economy composed of small businesses in vigorous competition has been a mirage at least since the Civil War. According to one observer, "In the age of late capitalism, with its overwhelming concentration of wealth and power, the vision of a *laissez-faire* society of individual entrepreneurs is nothing more than a handy myth for conservative ideologues."[2] In 1976 the largest 466 manufacturing corporations owned 71 percent of all manufacturing assets and earned 72 percent of all net (after tax) profits in that sector. In the economy as a whole the richest 0.2 percent of corporations (only 3,800 in 1974)

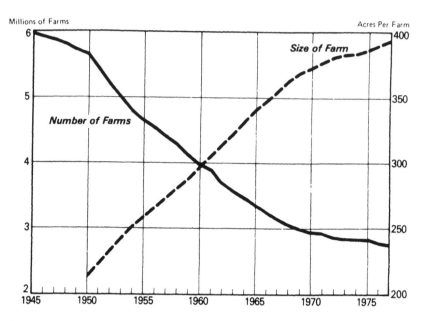

Figure 1.2. Number of farms and average size of farms, 1945–1977.

owned 72 percent of all corporate assets, produced 52 percent of sales and earned 69 percent of corporate profits.[3] Personal wealth is also concentrated: 2 percent of the households in this country own 80 percent of all individually held corporate bonds.[4]

Agriculture is distinctive for *resembling* the ideals of the competitive market economy more than almost any other sector of the U.S. economy does. Within U.S. agriculture the institution of the family farm is an anachronism of the national market economy, the competitive characteristics of which have withered away long ago. The concentration of economic and political power in agriculture, with the supportive intervention of the state, renders the free enterprise model of agriculture illusionary as well. Agribusiness—agricultural input and processing corporations—encouraged and assisted by the federal government, has eliminated the competitive nature of agriculture. Under the cover of supporting family farmers, agribusiness and federal policies are actually destroying family farms and replacing them with artificially created large-scale producers (Figure 1.3). Consequently, in 1974, 153,122 farms or 6.6 percent of all farms produced 53.8 percent of the total value of agricultural products.

The family farm is an important part of American folklore extending back to the eighteenth century. Nominally, farm families own the land they work themselves, manage their farms on a day-by-day and long-term basis, provide the working capital, and take the risks of harvests and markets. In contrast, large-scale, capital-intensive farms—often owned by absentee landlords and

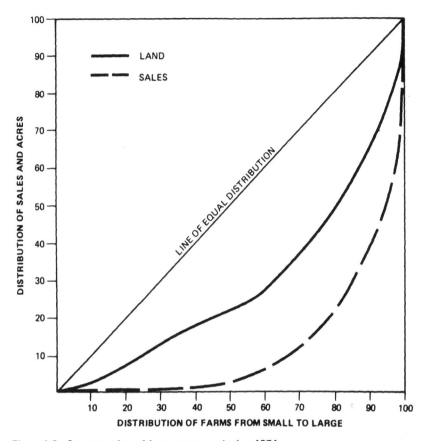

Figure 1.3. Concentration of farm acreage and sales, 1974.

operated by managers and farm workers—have traditionally been considered undesirable in American culture. Consequently, writers and politicians have exalted the virtues of the family farm. Indeed, the appeal of the family farm has become stronger as the number of working farms has decreased. Despite the loss of almost 50 million in the farm population since 1920, the Food and Agriculture Act of 1977 stated,

> Congress hereby specifically reaffirms the historical policy of the United States to foster and encourage the family farm system of agriculture in this country. Congress firmly believes that the maintenance of the family farm system is essential to the social well-being of the Nation and the competitive production of adequate supplies of food and fiber. Congress further believes that any significant expansion of non-family owned, large scale corporate enterprises will be detrimental to the national welfare.

The *ideal* of the family farm, however, has been turned into the *myth* of the family farm. According to the myth, family farms dominate agricultural production in almost all sections of the country. But who is served by the family farm myth? Herbert Schiller has outlined the purpose of myths in general: "Myths are used to dominate people. When they are inserted unobtrusively into popular consciousness, . . . their strength is great because most individuals remain unaware that they have been manipulated."[5] In a class society, myths serve to reinforce the strength of elite groups and to weaken the effectiveness of the vast majority. "By using myths which explain, justify, and sometimes even glamorize the prevailing conditions of existence, manipulators secure popular support for a social order that is not in the majority's long-term real interest. When manipulation is successful, alternative social arrangements remain unconsidered."[6] Myths serve useful functions for those who benefit by them and for those who want to assure that others think the way they do. Their strength and power of persuasion lies in their factual, historical aspects; their power to control people lies in their illusionary aspects. The conventional account of contemporary U.S. agriculture is widely accepted because it is based on the national ideal of the family farm derived from the Jeffersonian concept of agrarian democracy and from a small amount of truth. When a myth is widely accepted, this small amount of truth is perceived as the whole truth!

Farmers appear to be, and even believe themselves to be, independent, but in reality they are in the power of forces beyond their control. In the United States the myth of the family farm continues to be used for the benefit of a relatively small group of large-scale producers and agribusiness firms, while the vast majority of rural and urban people, believing the myth, pay the economic, political, social, and environmental costs of this fantasy. This book is written for this majority, who can benefit from an exposure of the fraudulent use of the family-farm concept. In demonstrating that agribusiness is served by the family-farm myth, the demystification of U.S. agriculture can allow family farmers, small-scale producers in general, and food consumers to see beyond the manipulations by agribusiness and an agribusiness-biased government and to decide the future of agriculture in their own self-interests.

Supporting Myths

The myth that family farms predominate in U.S. agriculture has been used to disguise the real trends of large-scale farming and agribusiness. Despite the ever-increasing size of farm units and greater integration of the farm sector with the corporate and oligopolistic national economy, the myth of the family farm has maintained its strength because it is supported by several other myths.

The Work Ethic Myth

Self-reliance and rugged individualism are characteristic of the conservative ideology of farmers.[7] Everyone who works hard can make good, and conversely, anyone who fails to make good is lazy or inefficient.[8] The work ethic in the rural value system gives farmers a feeling of merit for their own industriousness, but the mythical aspect of it is revealed by farmers' low incomes relative to other workers'. Family farmers work longer hours, have a greater variety of skills, and absorb greater weather and market risks than any other occupational group, yet their incomes have usually been half that of other workers. Only in 1972–1973 did the average farm income increase to 70 percent of an urban worker's income.

The Free Enterprise Myth

Theoretically, free enterprise in pursuit of profits without government interference results, according to conservative and agribusiness interests, in the lowest costs and highest-quality products at fair prices; in practice, agrarian free enterprise is a myth. In 1972 the U.S. government paid farmers $5 billion *not* to grow food, and this artificial scarcity increased the cost of food for consumers by $4.5 billion. In addition, the Federal Trade Commission estimated that the concentrated food industry extracted $2.6 billion in overcharges on seventeen major food lines alone. The United States does not have a free-market economy in agriculture. Federal farm programs determine how much is grown, what the support prices for agricultural products will be, which foreign farm products are permitted to enter the country in what quantities, and which farmers will get what income. Although free-enterprise advocates argue that such government "interference" causes agrarian problems, an alternative interpretation is that the state has intervened in an attempt to manage problems that have originated in the "free-market economy" itself. In doing so, the state takes the blame off the private sector for its own mismanagement: instead of people rebelling against business, they rebel against government programs. Responsibility for the farm crisis has been shifted from its origin, the private sector of agribusiness, to the public sector.

The myth of free enterprise also serves to blame the victims of agrarian capitalism. Competition among family farmers is viewed as necessary and indeed beneficial for achieving upward mobility and moving up the agricultural ladder from tenant to part-owner to full-owner status by reinvesting "profits" in land and equipment. According to the myth, farmers who do not survive in this economy fail because of their own shortcomings. In practice, competition among small-scale producers results in their exploitation—seemingly self-imposed—and their eventual destruction, allowing large-scale producers to monopolize an even greater share of U.S. agricultural production.[9]

The Efficiency Myth

According to this myth, farmers who survive the competition of the market economy must be the most efficient producers. However, a disproportionate amount of government farm program payments go to large-scale producers. Rather than basing these subsidies on yields per worker or per acre, they are based on total farm production or land owned. In effect, the state renders efficient family farmers noncompetitive. The efficiency myth legitimates farm expansion and increased profits under the disguise of productivity.

The Equal Opportunity Myth

In theory, individual producers compete as equals in the marketplace. Federal laws frequently treat farmers alike, but when farmers have unequal access to land, capital, and technical know-how, the outcome is predictably unequal. Although officially the United States (both government and corporations) espouses equality, this rhetoric shields Americans from the truth of inequality in opportunity.[10]

The Crisis in Farming and Values

The genuine goal of farm families and the espoused goal of the federal government—to keep family farmers in agriculture—is not being achieved. Although each year the number of farmers decreases, the remaining ones continue to face economic uncertainty. Regardless of how hard family farmers work and how efficient they become, they are not treated equally with large-scale producers under capitalism. The farm crisis is centered in this unequal treatment of family farmers in a society allegedly dedicated to justice and equality.

This crisis can be resolved in two ways. We must view the purpose of U.S. agriculture either as short-term profit maximization for large-scale producers or as the means of providing employment and food for human need in the short *and* long term without disproportionately burdening a few groups with the costs of such production. A good way to illustrate the differences in explaining the trends in U.S. agriculture is by the use of a puzzle. The object is to connect the nine dots shown with *four* straight lines *without* lifting your pen from the page.

.   .   .

.   .   .

.   .   .

The space outlined by the nine dots represents the explanatory framework of proponents of agribusiness. But the solution to this puzzle can be found only by going outside the nine dots:

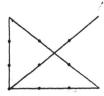

Likewise, a much broader frame of reference is required if we are ever going to adequately understand U.S. agriculture.

Another way of looking at this difference in perspective is the island analogy. A group of islands appears to be unconnected to each other; on the surface, each island is clearly separated from the next by water. Yet when the islands are examined from the ocean floor, they are connected. In agriculture, surface events—decreasing farm numbers, low farm incomes, "surplus" food production, and so on—can be adequately explained only by structures and relationships that are interconnected below the surface.

This book views the U.S. agrarian crisis from the perspective of family farmers and landless agricultural workers. It presents the view from the bottom, from the ocean floor, rather than the familiar view from the top, from the islands. The view from the bottom of society is more profound than that from the top because in a democracy the conditions and experiences of ordinary people, the majority, are more compelling than the views of the self-serving elite. The premise of this book is that farm workers who produce agricultural wealth in the form of food and fiber should reap the benefits of their labor and, with other workers, should decide how this wealth is to be distributed and for what purposes it is to be used. Such a view necessitates consideration of factors seldom presented in the standard analytical puzzle of the American farm crisis.

Overview of the Book

Family farming is a myth. How this myth is perpetuated and who benefits from the myth are the subjects of Chapters 2 through 11. Based on the conventional definition of a family farm, the U.S. Department of Agriculture's (USDA's) own data are used in Chapter 2 to show that family farms are far less important than the federal government claims. Two kinds of farm classifications are then compared. When other definitional characteristics such as indebtedness, off-farm incomes, and contract farming are included, the concept of independent family farmers is rendered meaningless.

Part 2 covers the consequences of early legislative attempts to achieve the family farm ideal. The federal government was the earliest exponent of agrarian democracy; secure control of land by yeoman farmers assured a politically free democracy. But, as Chapter 3 reveals, equal opportunity was rarely achieved in the many nineteenth- and early twentieth-century land grants. Even the well-known Homestead Act failed to achieve its objectives. Similarly, federal irrigation laws promised to establish family farming in the arid West, but instead agribusiness prospered. The consequences of inadequately written laws, mismanagement and abuses of the laws, and fraudulent transactions turned the ideals of family farmers into subsidies for land speculators and large-scale farmers. Water usage (Chapter 4) and landownership patterns (Chapter 5) today reflect the past agribusiness bias of the federal government.

In Part 3, the comfortable relationship between agribusiness and the market economy is discussed. Chapter 6 deals with farm size and the question of efficiency. In the United States, efficiency is equated with profitability. The arguments for productive efficiency have been purposely confused with large-scale farms and large profits. Although government studies indicate that a two-person farm is the most efficient farm unit, farmers expand beyond this size because profitability continues to increase. Farmers know, despite claims to the contrary, that profitability, not efficiency, is rewarded. Family farmers are efficient; but because they are not as profitable as large-scale producers, they go out of business and agribusiness strengthens its hold on the farm economy. Large-scale farms represent the production side of agribusiness. The characteristics of these farms, which dominate U.S. farm produce sales, are described in Chapter 7. The other components of agribusiness—farm input oligopolies and processing and marketing oligopsonies—control large-scale farmers and family farmers alike. Chapter 7 also indicates the ways agribusiness manipulates all farmers, regardless of size and type.

Part 4 examines the agribusiness bias of current federal government policies and programs. In the context of agrarian capitalism, federal programs reflect the pervasive economic norm of short-term profit maximization, which is camouflaged by rhetoric in support of family farming. Federal tax laws (Chapter 8), farm programs (Chapter 9), and agricultural research and extension work (Chapter 10) supposedly help family farmers but, like nineteenth-century land grants, actually give preferential advantages to large-scale farmers and discriminate against family farmers, particularly black farmers in the South.

The myth of the family farm has disguised the rural consequences of agribusiness, discussed in Part 5. The social, economic, and environmental costs of large-scale farming have been largely ignored by government agencies. Nevertheless, the plight of seasonal farm families (Chapter 11) and the decline of small towns dependent on family farms (Chapter 12) are the direct results of agrarian capitalism with its agribusiness and governmental manifestations.

In Part 6 the question of agrarian democracy versus agrarian capitalism is considered. Farmers have consistently challenged the invasion of agribusiness (Chapter 13). Although the Populists of the late nineteenth century demanded a radical restructuring of agriculture and of the U.S. economy, most subsequent farm organizations have had conservative goals, lashing out at symptoms rather than attempting to treat the root causes of the farm crisis.

Chapter 14 provides a theoretical framework in which to place the myth of the family farm and the reality of agribusiness. (Readers who want a theoretical perspective are encouraged to read this chapter first.) Because family farmers play multiple roles—landlords, capitalists, managers, and workers—the myth encourages them to identify with those roles that least threaten the status quo and consequently undermine their survival. The concept of surplus value and the labor theory of value provide the basis for a structural explanation of U.S. agriculture. Structurally, family farmers are essentially workers, but the family-farm myth teaches them to identify with agrarian capitalists. This false identification separates them from other workers and prevents them from working against their own exploitation to achieve equality and justice. Only when family farmers recognize that they produce agricultural wealth as workers, not as capitalists, and join in solidarity with other like-minded workers, can they achieve the goals of agrarian democracy.

What Is a Family Farm? Farm Definitions and Classifications

Scientists and politicians use the U.S. Department of Agriculture's data to support the myth of the family farm. But a critical review of this data indicates that family farmers do not predominate in U.S. agriculture. The family-farm myth rests on an inadequate data base and on an inaccurate interpretation of the data. Beyond the narrow definitional characteristics of family farmers are a series of additional factors, such as indebtedness, off-farm employment, and contract farming, that render the concept of the independent family farm largely illusionary.

How are family farms defined? Conventional definitions of farm structure include four characteristics: land, labor, capital, and management. Who owns the land, performs the labor, controls the capital, and makes the decisions determines the kind of farm operation. In this view little or no differentiation exists on family farms among who owns the farmland, buildings, machinery, and livestock; who performs the farm work; who provides the working capital; and who makes the day-to-day and long-term decisions. Family-farm production units are necessarily limited to a size that can be worked and managed by the farm family itself. Regardless of the particular kind of farming, the larger the acreage under cultivation, the greater the reliance on hired labor, and hence the less likely that the operations are family farms. Family farms, being owner operated, also require a close geographical association between household residence and farm operation, although the location of the farm and the place of family residence need not be identical. On the Great Plains, for example, farm families often live in towns but work their land in the adjacent countryside. Sidewalk farming, as this kind of farming is called, however, is more a result of the dry-farming techniques used in the arid West and the subsequent low population densities than of absentee, non-family farms.

In contrast, land, labor, capital, and management on industrial farms are controlled by separate sets of people. Farms owned by absentee individuals or corporations are operated on a daily basis by paid managers who supervise hired agricultural laborers. Those who manage and work industrial farms live on the land, but the owners usually live in cities. Such absentee ownership is more common in some types of farming than in others. In 1974, 19 percent of all farm owners lived off their farms, but 35 percent of cotton and 30 percent of fruit and nut farm owners had off-farm residences.[1] The heavy reliance on hired labor also allows industrial farmers to work large acreages and to produce large agricultural sales.

Evidence of Family-Farm Dominance

The USDA is *the* source of information and data on the structure of farm sizes, and, therefore, its data and analyses have tremendous influence in Congress and with scientists, farmers, and the public. What the USDA says about family farms determines how most professional and lay people perceive them. According to the USDA, recent data indicate:

1. Family farms currently dominate total farm numbers.
2. Family farms dominate food and fiber production and sales.
3. Relative to the total number and sales of U.S. farms, the proportion of family farms is not decreasing.
4. Incorporated farms are an insignificant percentage of total farm numbers and sales.

The USDA uses four separate sources of data to support these conclusions.

Classification of Farms by Amount of Labor Hired

Radoje Nikolitch, an agricultural economist for the USDA, has conducted research showing that family farms predominate in the U.S. farm economy. Based on a special tabulation of the Census of Agriculture, he divided all farms into two categories: "Family-managed farms that use less than 1.5 man-years of hired labor are considered to be family-operated farm businesses [*family farms*] ; those that use more than 1.5 man-years of hired labor or have hired management are *larger-than-family farms*."[2] He concluded that family farms accounted for 95 percent of all farms and about 63 percent of all farm sales in 1949 and 1969. Larger-than-family farms represented only 5 percent of all farms and about 38 percent of all farm sales in this twenty-year period (see Table 2.1). The importance of family farms varied with the size of farms. "In 1964, family farms accounted for only 14.5 percent of the number and 8.4 percent of the sales of farm products of farms with sales of $100,000 or more. But they accounted for 99.8 percent of the farm numbers and 99.7 percent of farm product sales on farms with less than $5,000 of sales."[3]

Family farms predominate in highly mechanized types of farming, according to Nikolitch. "In 1964, about 97 percent of all cash-grain farms were family units, and they accounted for 85 percent of total annual sales by all cash-grain farms."[4] In contrast, family farms were less important for fruit and vegetable farming, which are the least mechanized. "In 1964, family farms accounted for 83 percent of all vegetable farms, but only 15 percent of total annual sales by all vegetable farms. For fruit and nut farms, the proportions were 83 percent and 29 percent"[5] (see Table 2.2).

TABLE 2.1
Number of Farms and Sales of Farm Products by Class of Farms, 1949-1969[a]

Class of Farm	Number of Farms[b]				Sales of Farm Products[b]			
	1949[c]	1959	1964	1969[d]	1949[c]	1959	1964	1969[d]
	1,000 Farms				Million Dollars			
All Farms	4,905	3,695	3,150	2,726	22,280	30,362	35,075	44,026
Larger-than-Family Farms	264	165	154	146	8,250	9,226	12,427	16,730
Family Farms	4,641	3,530	2,996	2,580	14,030	21,136	22,648	27,296
Less than $10,000 Sales	4,301	2,892	2,271	1,773	9,282	8,391	6,413	6,164
$10,000 or more Sales	340	638	725	807	4,748	12,745	16,235	21,132
	Percent of All Farms				Percent of All Sales			
All Farms	100	100	100	100	100	100	100	100
Larger-than-Family Farms	5	5	5	5	37	30	35	38
Family Farms	95	95	95	95	63	70	65	62
Less than $10,000 Sales	88	78	72	65	42	28	18	14
$10,000 or more Sales	77	17	23	30	21	42	47	48

[a]Any farm not operated by a paid manager and using less than 1.5 man-years of hired labor, regardless of any other characteristics, is classified as a family farm. Any farm using 1.5 or more man-years of hired labor operated by a paid manager is classified as a larger-than-family farm.
[b]Institutional farms and farms in Alaska and Hawaii not included.
[c]1949 corrected for change in farm definition in 1959. Sharecropper operations not considered as independent farms but as parts of respective multiple-unit operations.
[d]Estimated by projecting census information.

Source: Data derived from special tabulations by the Bureau of the Census from 1959 and 1964 Censuses of Agriculture cited in Radoje Nikolitch, Family-Size Farms in U.S. Agriculture, USDA, Economic Research Service Report No. 499 (Washington, D.C.: Government Printing Office, 1972), Table 1.

TABLE 2.2
Family Farms and Family Farm Sales as a Percentage of All Farms by Type, 1959-1964

Type of Farm	Family Farms as a Percentage of All Farms[a]		Percentage of Total Farm Products Sold by Family Farms, 1964[b]
	1959	1964	
Cash grain	98	97	85
Tobacco	98	97	82
Cotton	92	90	42
Other Field Crop	86	80	30
Vegetable	86	83	15
Fruit and Nut	88	83	29
Poultry	96	92	57
Dairy	95	94	77
Other Livestock (other than poultry and dairy)[c]	97	97	69
General	95	95	70
Miscellaneous	86	96	29
Total	95	95	65

[a]Institutional farms and farms in Alaska and Hawaii not included, except farms in Alaska with $100,000 or more of farm products sold.
[b]Estimated on the basis of special tabulations from the Bureau of the Census for 1959 and 1964.
[c]Includes livestock ranges.

Source: Radoje Nikolitch, Family-Size Farms in U.S. Agriculture, USDA, Economic Research Service Report No. 499 (Washington, D.C.: Government Printing Office, 1972), Table 2.

He also found that family farms predominated in every region of the United States in both 1959 and 1969. They "accounted for 98 percent of all farms in the Lake States, Corn Belt, and Northern Region. The percentage was smallest in the Pacific Region—87 percent of all farms in both years. In the other regions, the percentage ranged from 89 to 96 percent, with little significant change from 1959 to 1964."[6] In the rest of the country family farms had a slight predominance. In contrast, larger-than-family farms predominated only in the Southwest, Florida, and in a few widely scattered states. Figure 2.1 shows the predominance of family-farm sales by region.

Nikolitch explained this concentration of nonfamily farms in two ways. First,

In some types of farms, obtaining inputs, marketing, and processing have become so closely related to production that their organization is being readjusted to best coordinate these functions. . . . As a result, these farms tend to develop into larger business sizes, and the patterns of organization found in nonfarm industries are becoming more prevalent. This is true of farms producing vegetables, fruits, eggs, and feed cattle—the leading products of California, Texas, Florida, and a few other states where family farms are less dominant. This suggests that regional differences are partly related to type of farming.[7]

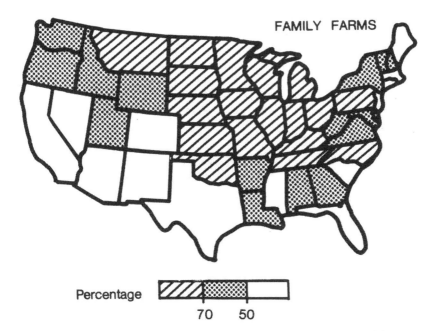

Figure 2.1. Farm sales by family farms as a percentage of total sales, 1964. *Source:* Radoje Nikolitch, *Family-Size Farms in U.S. Agriculture,* USDA, Economic Research Service Report No. 499 (Washington, D.C.: Government Printing Office, 1972).

Second,

There are also historical differences in the development of the structural organization of farming in these various regions. The family pattern of organization of the first colonists in the East and the "Homestead" family settlement of the Midwest still influence these parts of the country. In the same way, the plantation system in the South and the Spanish latifundia heritage in California, Arizona, New Mexico, Texas, and Florida may still be factors in these areas.[8]

Classification of Farms by Land Tenure

Land-tenure data from the Census of Agriculture is a second source of evidence used to demonstrate that family farms predominate in the United States. The four categories are: (1) full owners—all land worked is owned by operators; (2) part owners—only part of the land worked is owned by operators; (3) tenants—most of the land worked is rented by operators; and (4) hired managers—operators neither own nor rent land but are paid directly by owners. Presently, owner-operator farms (full and part) occupy a predominant position in terms of numbers of farms and acreage (see Table 2.3).

Classification of Farms by Type of Orgnaization

The USDA also classifies farms by their legal status: sole proprietorship (individual or family), partnership, and corporation (by ten or fewer shareholders and by more than ten shareholders). Farms owned by an individual or family and farms held by partnerships would fall into the category of family-owned farms. Farm corporations with ten or fewer shareholders under most circumstances would also qualify as family farms, according to the

TABLE 2.3
Farms and Land in Farms by Tenure of Operator, 1900-1974

Year	Farms		Land	
	Owner-Operated	Nonowner-Operated	Owner-Operated	Nonowner-Operated
	(Percent)		(Percent)	
1900	63.7	36.3	66.3	33.7
1920	60.9	39.1	66.7	33.3
1940	60.7	39.3	64.3	35.7
1964	82.4	17.6	76.7	23.3
1969	88.1	11.9	88.9	13.1
1974	88.7	11.3	88.1	11.9

Source: Census of Agriculture.

USDA. Table 2.4 clearly shows that family farms predominate in numbers and in sales. This data would seem to confirm the USDA's conclusion that "sales from all corporate farms accounted for about eight percent of total sales of all farms. Sales from nonfamily corporations were about two percent of total sales."[9]

Composition of the Farm Work Force

Because of the rapid substitution of capital—in the form of harvesting machines, artificial fertilizers, and pesticides—labor used on farms has declined. Since the end of World War II, about 70 percent less labor has been used to produce 36 percent more food and fiber, yet family labor is predominant on a majority of U.S. farms (see Table 2.5). "The crucial question with regard to family farms is whether this substitution of capital for labor has displaced more family labor or more hired labor."[10] Nikolitch concluded that "the long-term predominance of self-employment in American farming has not been lessened significantly by recent economic, social, and technological changes. . . . In the country as a whole, family labor accounted for 76 percent of total farm employment during 1930–39 and 75 percent in 1967."[11]

Evidence of Large-Scale Farm Dominance

Richard Rodefeld, an academic agricultural economist, has made pointed criticisms of all four sources of data and the arguments employed by the USDA, raising questions about the data's reliability, completeness, and interpretation. His arguments are significant because they show that even a conventional critique of the USDA data, without an examination of the questions underlying the data, indicates that official statements on the predominance of family farms are either questionable or incorrect. Rodefeld attacked each of the four sources of data and the arguments upon which they are based and found that the USDA's own data in fact indicate reverse trends.

Classification of Farms by Amount of Labor Hired

Rodefeld argued that Nikolitch excluded information that is inconsistent with the conclusion that family farms predominate in U.S. agriculture. Although little change occurred in the proportional numbers of family farms and larger-than-family farms from 1959 to 1964, the proportion of farm sales did change during this period. As a percentage of all farm sales, family farms accounted for 70 percent in 1959, 65 percent in 1964, and 62 percent in 1969 (see Table 2.1).

According to Nikolitch, from 1949 to 1959 the number of larger-than-family farms dropped by 99,000 farms or a 37 percent decrease, but family-sized farms declined by only 24 percent. Correspondingly, the proportion of farm sales increased only 12 percent for other-than-family farms but increased by 51 percent for family farms. But Rodefeld said, "This is a curious situation because conventional wisdom and accumulated knowledge suggest that smaller farms have higher mortality rates than larger farms."[12] The explanation

TABLE 2.4
Legal Structure of Businesses Reporting Farm Income, 1958-1968

Legal Structure	Number of Businesses		Business Receipts		Average per Tax Return[a]	
	1958	1968	1958	1968	1958	1968
	Thousands		Billion Dollars		Dollars	
Sole Proprietors	3,343.2	3,042.6	22.4	35.0	6,705	11,503
Partnerships	136.6[b]	109.9	3.4[b]	4.7	25,080	42,766
Corporations	8.7	20.0	2.2[b]	5.7	249,281	285,000
Total	3,488.5	3,172.5	28.0	45.4	8,028	14,310
	Percent					
Sole Proprietors	96	96	80	77	---	---
Partnerships	4	3	12	10	---	---
Corporations	0[c]	1	8	13	---	---
Total	100	100	100	100	---	---

[a]Computed from unrounded data.
[b]Estimated for 1958.
[c]Less than 0.5 percent.

Source: Internal Revenue Service, unpublished data, cited in Radoje Nikolitch, Family-Size Farms in U.S. Agriculture, USDA, Economic Research Service Report No. 499 (Washington, D.C.: Government Printing Office, 1972), Table 15.

TABLE 2.5
Percentage of Farms and Farm Sales by Man-Years of Hired Labor, by Class of Farms, 1964

Man-Years of Hired Labor	Total		Large[a]		Medium[b]		Small[c]		Part-Time[d]	
	Number	Sales	Number	Sales	Number	Sales	Number	Sales	Number	Sales
Family Farms										
None	49.8	19.4	2.1	1.3	15.7	13.6	35.0	32.6	65.9	52.0
Less than 0.5	38.0	30.4	3.9	2.2	35.5	30.6	50.3	50.4	32.5	40.5
0.5-0.9	4.9	8.2	3.3	1.9	13.3	12.6	8.5	9.4	1.2	2.3
1.0-1.4	2.6	6.5	5.7	3.3	10.9	11.4	3.5	4.2	0.3	0.6
Larger-than-Family Farms										
1.5-2.4	2.1	7.3	10.5	6.3	11.3	12.9	2.0	2.5	0.1	0.4
2.5-4.9	1.5	8.2	20.7	13.7	9.1	12.0	0.6	0.8		0.8
5.0-6.9	0.4	3.3	11.2	8.2	2.2	3.3	0.1	0.1		0.5
7.0-9.9	0.3	2.9	10.8	8.8	1.2	2.0	e			0.5
10.0-14.9	0.2	3.1	11.0	10.9	0.6	1.1				0.5
15.0 and over	0.2	10.7	20.8	43.4	0.2	0.5				1.9
All farms	100.0	100.0	100.0	100.0	100.0	100.0	100.0	100.0	100.0	100.0

Percentage Distribution by Farm Size

[a]Farms with sales of $100,000 or more.
[b]Farms with sales of $20,000 to $99,999 inclusive.
[c]Farms with sales of $5,000 to $19,999 inclusive.
[d]Farms with sales of $5,000 or less.
[e]Blank spaces indicate less than 0.05 percent.

Source: Special tabulation, 1964 Census of Agriculture, cited in Radoje Nikolitch, "Family-Operated Farms: Their Compatibility with Technological Advance," American Journal of Agricultural Economics 51 (1969), p. 537.

for this seemingly contradictory trend is found in a footnote in Nikolitch's study. For the 1949 data, Nikolitch did not consider sharecropper operations as independent farms but as parts of multiple-unit plantations (see Table 2.1). This procedure inflated the number of larger-than-family farms in 1949 in comparison with the 1959–1969 data. "In 1959 and later, however, 'multiple-unit operations' were not counted as single units but instead all the sharecropper units on these places were counted separately. Hence, larger-than-family farms experienced a big drop from 1949 to 1959."[13]

If consistent definitions of farms had been used in all years, the conclusions reached by Nikolitch and the USDA actually would have been the reverse of those reported. As Rodefeld pointed out, "This data, instead of being evidence for family farm stability, is in actuality evidence of systematic, linear family farm decline and larger-than-family size farm growth."[14] By regrouping the data presented by Nikolitch, larger-than-family farms would either predominate or be close to it in most regions and states, except in the Midwest. This is also true of most types of production other than those concentrated in the Midwest (see Table 2.6).[15]

Classifying farms by the amount of labor hired excludes other critical features of family farms. Specifically, landownership and ownership of non-land resources need to be considered in any comprehensive classification of farm sizes. Tenant farms and family farms may rely mostly on family labor, but their control over land resources distinguishes them from each other. Nikolitch's classification fails to differentiate tenant from family farms; yet if he had considered this important distinction, the owner-operated family-

TABLE 2.6
Percentage of Larger-than-Family Farm Numbers and Sales, by Region and Type of Production, 1964

Region	Percentage Farms	Percentage Sales	Type of Production	Percentage Farms	Percentage Sales
Pacific	13	71	Vegetables	17	85
Southeast	6	56	Fruit and Nuts	17	71
Mountain	9	54	Other Field Crops	20	70
Delta	8	51	Cotton	10	58
New England	11	49	Poultry	8	43
Southern Plains	6	45	Miscellaneous	5	35
Middle Atlantic	7	35	Other Livestock	3	31
Appalachia	4	27	General	5	30
Northern Plains	2	18	Dairy	6	23
Lake States	2	14	Tobacco	3	18
Corn Belt	2	13	Cash Grain	3	15
United States	5	35	Total	5	35

Source: Radoje Nikolitch, Family-Size Farms in U.S. Agriculture, USDA, Economic Research Service Report No. 499 (Washington, D.C.: Government Printing Office, 1972), pp. 7 and 9.

sized farms would have already accounted for less than 50 percent of all farm sales by 1959.[16]

Classification of Farms by Land Tenure

Rodefeld also suggested that "some serious questions must be raised about the utility and meaningfulness of relying on land tenure data,"[17] as defined by the census, to assess the nature of family farming. For example, the percentage of farm acreage operated by full owners declined from 51 percent in 1900 to 29 percent in 1964, but increased to 35 percent in 1969. This increase was not related to what was actually happening on family farms, but was the consequence of the USDA's redefinition of the term "farm operator." Prior to 1969 the farm operator was the person managing the farm operation on a daily basis, doing the farm work, or directly supervising such work; in 1969 the farm operator became "the person in charge of the farm or ranch operation." "In charge" meant the ability to control, by virtue of proprietorship alone, the organization and operation of farms. Farm operators were thus no longer restricted to dirt farmers but also included corporations, such as Del Monte, and wealthy absentee Wall Street farm owners. In addition, farms operated by hired managers completely vanished.[18] Furthermore, land-tenure data, like almost all other data from the Census of Agriculture, cannot be compared with other critical farm characteristics. By dealing only with ownership status, whether owner- or nonowner-operated farms, the amount of hired labor and the ownership of other-than-production resources are ignored (see Figure 2.2).

Classification of Farms by Type of Organization

The Census of Agriculture provides data on the legal organization of farms but fails to cross-reference these data with other farm features. Although used to demonstrate the predominance of family farms, a simple legal definition of ownership allows for some types to be placed in the family category when they effectively belong in the corporate category. According to Rodefeld, "Family-owned farms which would under no circumstances be classified as 'family' farms are: absentee family-owned farms with hired managers and hired workers; and, family and managed farms with high inputs of hired labor."[19] Using Rodefeld's structural rather than legal definitions, corporate farms (1) are large, in terms of acreage and/or farm sales, (2) are owned by absentee owners, (3) are operated by paid managers who use hired labor, and (4) provide their own capital. Yet the USDA's definition of corporate farms considers only those legally incorporated and the number of shareholders in the corporation.

> [The fact] that the USDA's definition of a "corporate" farm may be excluding the majority of structurally defined "corporate" farms is indicated by research I was involved with at Wisconsin, where it was found that approximately 75 percent of all structurally defined legally incorporated "corporate" farms were owned by an individual, family, or small group of unrelated individuals. None of these "corporate" farms would have been included in the USDA's "corporate" farm cate-

NUMBER OF FARMS, BY TENURE OF OPERATOR

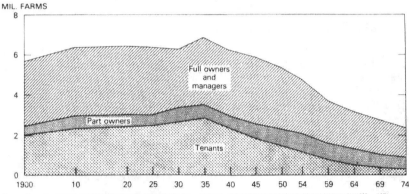

MIL. FARMS

Managed farms were discontinued in 1969. Such farms were classified by tenure based on whether the land operated was owned or rented after 1969. 1974 preliminary data. Source: Census of Agriculture.

LAND IN FARMS, BY TENURE OF OPERATOR

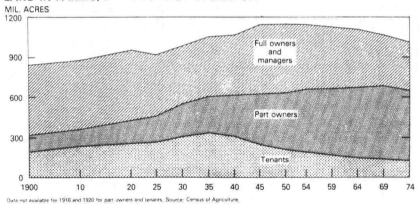

MIL. ACRES

Data not available for 1910 and 1920 for part owners and tenants. Source: Census of Agriculture.

Figure 2.2. Number of farms and acres in farms by land tenure, 1900–1974.

gory. . . . [The] "corporate" farm category undoubtedly is even greater than 75 percent because no assessment was possible on the number of such farms unincorporated in Wisconsin.[20]

Composition of the Farm Work Force

The USDA claims that if the proportion of family labor to total farm labor is not declining, then family farms are not declining. Despite the fact that family-farm labor remained the same proportion (96 percent) of total farm labor from 1959 to 1964, family-farm labor produced a lower percentage (a drop of 13 percent) of the total value of farm products sold.[21] In other words, although the percentage of larger-than-family farms remained the same from 1959 to 1964, their proportion of the total value of farm products sold increased by 13 percent.

Alternative Farm Classifications

The major problem with the USDA's classification categories is that they consider only one factor at a time. An accurate identification of family farmers requires the simultaneous inclusion of at least two characteristics. For example, the use of family labor alone would indicate a family farm. But if that farm is owned by an absentee landlord and operated by sharecroppers, it can hardly be used to support the agrarian ideal.

Land Tenure and Labor Classification

Rodefeld proposed a classification scheme with four structural categories based on land tenure *and* sources of labor (see Table 2.7).[22] *Family-sized farms* are completely or mostly owned, managed, and worked by farm families. They represent the ideal family-farm type that has been the dominant model for farming in the United States for the last two hundred years. *Industrial farms* with their absentee ownership and paid managers supervising hired workers lie at the other end of the farming continuum. They are commonly associated with incorporated farms and include plantations when operated as single farm operations. Two other types of farms occupy positions between these two extremes. *Larger-than-family farms* are owner operated (over half of the farmland is owned by the operator), but the bulk of the physical labor is done by hired workers. Most fresh vegetable and fruit farms, livestock ranches, and some plantations are classified in this category. *Tenant farms* are characterized by absentee ownership, with the tenant family own-

TABLE 2.7
Farm Types Defined by Sources of Labor and Types of Tenure

Sources of Labor	Types of Tenure	
	Owner-Operator	Nonowner-Operator
Mostly Family	Family Farm	Tenant Farm
Mostly Hired	Larger-than-Family Farm	Industrial Farm

Source: Richard D. Rodefeld, "A Reassessment of the Status and Trends in 'Family' and 'Corporate' Farms in U.S. Society," *Congressional Record*, 93d Cong., 1st sess. (May 31, 1973), p. S10059.

Figure 2.3. Tenants and sharecroppers displaced by agribusiness. (Top) Tenant farmers displaced by tractors, Texas, 1937. Dorothea Lange, FSA photo. (Bottom) Black sharecroppers evicted from plantation farms, New Madrid County, Missouri, January 1937. Arnold Rothstein, FSA photo.

ing less than 50 percent of the farmland and providing most of the labor and management for the operation. All family-sized farms, rented or leased, and sharecropper farms are included in the tenant category. Sharecroppers have less management control over the farm operation (which crops are grown and when they are planted, sprayed, and harvested) and provide less capital because landlords supply seeds, fertilizers, and machinery in return for one-third to one-half of the crops.

Incorporated farms can be found among all four farm types. Many family farms and larger-than-family farms are legally incorporated for the benefits of intergenerational farm transfers and federal and state income and inheritance laws. Corporations that engage in farming as part of their vertical integration (controlling aspects of production, transportation, processing, and marketing) are found only in the industrial farm category. These particular industrial farms are also incorporated, but they do not necessarily have to be incorporated industrial farms. The legal description of a farm is often different from its operational characteristics. (Corporate farms are discussed in greater detail in Chapter 7).

Using Rodefeld's classification, what does the USDA data indicate about family farms? Tenant and industrial farms are automatically excluded from the family-farm category. Larger-than-family farms are also excluded because hired labor does most of the work, even though decision making and ownership of resources are still largely in the hands of farm families. This leaves family farmers, who accounted for 79 percent of all farmers but only 49 percent of all agricultural sales in 1964. The USDA's unidimensional classification indicated 95 percent and 63 percent, respectively. Conversely, larger-than-family and industrial farmers together, representing 5 percent, produced 35 percent of all farm sales (see Table 2.8). Although the proportion of family-farms of the total number of farms and their share of farm produce value has changed relatively little, industrial farms have significantly increased their share of the value of farm products sold, especially given the small number of such farms.

Even though Rodefeld's classification helps to reveal the increasing significance of large-scale farming, Table 2.8 overestimates the importance of family farms and underestimates larger-than-family and industrial farms in two ways. First, as they are based on USDA surveys, the number and sales of industrial farms are extremely conservative estimates. If plantations instead of individual sharecropper units were enumerated, the importance of tenant-farm numbers and sales would be sharply reduced and the importance of larger-than-family and industrial farms would be increased. Second, the importance of family farms is extremely inflated by including all farms, both "commercial" and "noncommercial," in the calculations. Noncommercial farms are those that sell less than $2,000 of farm produce per year. In 1964, they comprised 41 percent of total farm numbers but contributed only 3 percent of sales. Nearly all of them are family and tenant farm types, but their nominal contribution to the agricultural economy provides little support for family-farm viability. Exclusion of such subsistence and hobby farming would greatly increase the relative proportions of larger-than-family and

TABLE 2.8
Farm Types by Numbers and Sales, 1959-1964

Farm Type	Farms (Thousands)			Sales (Millions)			Percent of All Farms		Percent of All Sales	
	1959	1964	Percent Change 1959-64	1959	1964	Percent Change 1959-64	1959	1964	1959	1964
All Farms	3,695	3,150	-14.7	$30,362	$35,075	+$15.5	100.0	100.0	100.0	100.0
Family	2,808	2,475	-11.9	15,224	17,276	+ 11.9	76.0	78.6	50.1	49.3
Tenant	722	521	-20.7	5,912	5,372	- 9.0	19.5	16.5	19.5	15.3
Larger-than-Family	139	122	-12.2	7,202	8,915	+ 20.8	3.8	3.9	23.7	25.4
Industrial	26	32	+23.1	2,024	3,512	+ 73.5	0.7	1.0	6.7	10.0

Source: See Table 2.7.

industrial farms. The myth of the family farm thus rests on artificially inflated numbers.

Classification by Scale of Production

Rodefeld's two-variable classification is a major improvement over the USDA's single-variable definition of farm types. But a classification based solely on types of land tenure and sources of labor is still inadequate because it ignores the importance of gross farm sales and acreage operated.[23] The size of farm sales and acreage of individual farms affect local communities and the national economy. Farms with large farm sales can dominate the production of food and fiber regardless of how many smaller family farms exist. These large-scale farms are closely tied to large agricultural supply and food-processing and -marketing companies. In fact, *farms with large farm sales constitute the production side of agribusiness.* When farmers of such operations also control large acreages through ownership and/or leases, the total number of farm families in a rural community is reduced and small towns are in danger of losing their economic viability (see Chapter 12).

An ideal classification system that would render the most accurate definitions of farm types would have to *simultaneously* consider land tenure, labor hired, acreage, and gross farm sales. Unfortunately, such statistical cross-tabulation data is not available from the U.S. Census of Agriculture.[24] We are thus forced to present only a partial definition by choosing among the criterion variables. Thus restricted, the most influential characteristic, structurally, is gross farm sales (Figures 2.4 and 2.5). the value of farm production is far more important in understanding the structure and trends of U.S. agriculture than the number, size, and legal status of farms. Indeed, the myth of the family farm rests on the lack of distinction among these characteristics. For example, a dairy farm may have 150 cows and operate 480 acres, whereas a cattle ranch may have 1,000 animals and work 2,500 acres. Although the sizes vary, each operation is large scale because it produces a large value of gross farm sales.

Using unpublished 1974 U.S. Census data, a study by the General Accounting Office provides the basis for a scale-of-production classification.[25] *Large-scale* farms produce $100,000 or more of agricultural products. In 1974, they represented 7 percent of all farms and produced 54 percent of total value of sales. *Medium-scale* farms, with sales of from $99,999 to $40,000, represented 14 percent of all farms and 24 percent of farm sales. *Small-scale* farms, with sales of less than $40,000, represented 79 percent of all farms and produced 22 percent of all farm sales. The percentage of farmland operated also varied greatly between each size category. In 1974, large-scale farmers operated 27 percent of all farmland or had an average size of 2,064 acres; medium-scale farmers operated 24 percent of all farmland or had an average farm size of 761 acres; and small-scale farmers operated 49 percent of all farmland or had an average farm size of 310 acres. In this book, the terms large-scale and small-scale are used as defined here and the terms, respectively, stand for the production side of agribusiness and for family farms (see Table 2.9). (Medium-scale farms are rarely discussed here because they share

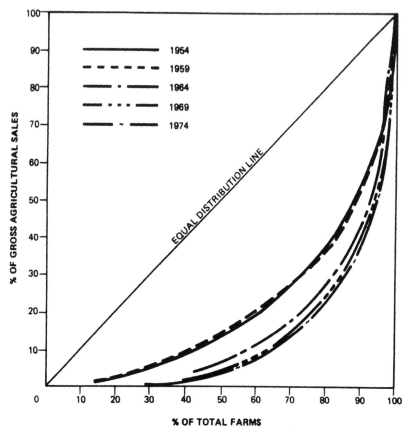

Figure 2.4. Concentration of farm sales, 1954-1974.

TABLE 2.9
Characteristics of Large-Scale and Small-Scale Farms

Characteristic	Large-Scale Farms	Small-Scale Farms
Land	Owned/Rented	Mainly Owned
Technology	Capital-intensive	Labor-intensive
Organization	Bureaucratic	Family Organized
Capital	Abundant	Scarce
Hours of Work	Regular	Irregular
Hired Labor	Important	Unimportant
Fixed Costs	Large	Small
Government Aid	Large	Small
Tax Laws	Beneficial	Harmful
Agricultural Research	Initiator	Receiver
Yields per Acre	Low	High
Yields per Worker	High	Low

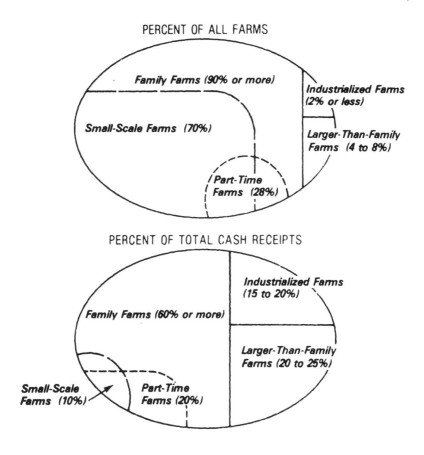

Figure 2.5. Types of farms: numbers and cash receipts in 1977. *Family farms* are farms that use less than 1.5 man-years of hired labor and are not operated by hired managers. *Industrialized farms* use assembly-line production techniques and their capital, owner-ship, management, and labor are highly differentiated. *Larger-than-family farms* are non-industrialized farms that use more than 1.5 man-years of hired labor. Farms can also be classified and defined according to their annual amount of sales or the annual number of days operators are employed off farms. *Small-scale farms* are farms with annual gross sales of less than $20,000; *part-time farms* are farms whose operators are employed off-farm 200 or more days per year. The precise overlap between these two types of farms and the other types is not known.

characteristics of each group—they are becoming either larger or smaller. In a polarized agriculture dominated by agribusiness, medium-scale farms are difficult to maintain.)

In comparison to Rodefeld's classification, large-scale farms include larger-than-family and industrial farms, and small-scale farms consist of family and tenant farms. Because of economic pressures and incentives, tenants and sharecroppers aspire to become family farmers and larger-than-family farmers are en route to becoming industrial farmers, although most would vehemently deny such a suggestion. Larger-than-family farmers are in an especially awkward position because they try to uphold their image as family farmers while taking on more and more of the characteristics of industrial farmers. Contrary to popular opinion, larger-than-family farmers, not industrial farmers, have been gobbling up family farms since 1964. Under current market conditions and federal farm policies, larger-than-family farms are the most viable; hence, family farmers are forced to expand their scale of production and acreage to survive. These large-scale farms are more capital-intensive and less land-intensive than small-scale farms, which are more labor-intensive. That is, small-scale producers have higher yields per acre than large-scale farmers.

Other Definitional Considerations

No one denies the rapid decline in the number of farms in the United States since World War II, or the rapid increase in farm size in all farm types. Conventional analyses of USDA data, including Rodefeld's, seek to answer the most-often-asked question: Are more small family farms being replaced by adequately sized family farms than by larger-than-family farms?[26] This question is concerned less with why small-scale family farms are disappearing than with how much labor and management are supplied by those who are still in business. This kind of analysis maintains the family-farm myth because it measures the decline of an institution that is assumed to still exist. The myth of the family farm relies heavily on a limited *descriptive* definition. Analyzing social change within traditionally accepted parameters is similar to the usual attempted solution to the nine-dot puzzle in Chapter 1. Understanding why family farming is declining and the current state of its existence requires a look beyond the variables considered by the USDA and conventional academics to seldom considered factors. If the ideal family farmer controls land, capital, labor, and management decisions, the degree of that control cannot be assessed until we know how family farmers relate to the overall structure of agriculture and to the national economy as a whole. Just as islands appear independent of each other but are interconnected below the surface, family farmers are connected to the economic system by indebtedness, off-farm incomes, and contract farming. If these factors are included in the search for real family farmers, one must question whether any exist who remain independent entrepreneurs, and thus the myth of the family farm is destroyed.

Capital and Decision Making: Indebtedness

The availability and cost of capital to farmers affect their ability to operate family farms. With increasing farm size and mechanization, off-farm companies supply a greater share of fixed and variable capital. The traditional definition and role of family farmers, which include the control of capital, are eroded as farm indebtedness increases. Unlike borrowing money from banks to buy houses for shelter, farmers borrow money to operate businesses.

The continual production of crops and livestock necessitates a continuous flow of money between farmers and bankers. Farmers pay back loans and then take out new ones to meet new expenses such as the purchase of larger equipment and more land. Since this farmer-banker interaction is unequal (after all, farmers cannot deny bankers loans!), bankers gain control over important decisions formerly made by farmers. For example, bankers deny dairy farmers loans if they do not have an "adequate" number of milking cows. Bankers make their own agricultural decisions—based on the ability of farmers to pay back loans—regardless of what individual farm families might want to risk. Bankers are also known to make conservative decisions to safeguard their investments. But what makes sense to bankers often makes little sense to farmers. What Lawrence Goodwyn, an economic historian, said about the U.S. banking system in general applies to agriculture as well. "The price of banker control of the nation's monetary system over the last 100 years has been incalculable in terms of the cramped lives of hundreds of millions of Americans. During these many decades, the private banking system has been the central pivot in the centralization of American life."[27] Nevertheless, as long as the economy remains structured as it is today, family farmers will continue to lose control over major and long-term farm production decisions.

How indebted are U.S. farmers? In 1974, 30 percent of farm operators had real estate debts and 24 percent had other farm debts. The total indebtedness of U.S. farmers was $33.7 billion, or an average of $71,726 per indebted farm. Farm real estate debt as a percentage of farm real estate assets was 11 percent in 1977, up from 8 percent in 1955. But as debts must be paid from current income and not from inflated land values, a more realistic measure of indebtedness would be related to income. Farm real estate debt as a percentage of all family income was 80 percent in 1975, up from 58 percent in 1960. When only income from farm sources is considered, the debt was 159 percent in 1975, up from 96 percent in 1960.[28] Because real estate indebtedness was 55 percent of *all* farm indebtedness in 1977, the ratio of total debt to income must be nearly doubled: total farm indebtedness represented 145 percent of all income and 289 percent of farm income! Even conservative agricultural economists, such as B. Delworth Gardner and Rulon D. Pope, recognized that "in percentage terms, increases in debt since 1970 are not matched by increases in farm family income. If debt continues to rise at a faster rate than net income, a critical liquidity problem as well as a net income problem could soon arise."[29] High indebtedness and the price-cost squeeze, which result from farmers receiving lower prices for their products than they pay for supplies, are forcing farmers to rely increasingly on banks to maintain their financial solvency or face bankruptcy. The independence of family farm management is reduced by greater dependence on bankers.

Off-Farm Incomes

The increasing amounts of off-farm incomes provide another sign that family farming, as a way of producing an adequate family income, is a myth (see Figure 2.6). Multiple-job holders characteristically receive inadequate incomes from any one job to meet their living needs and to pay off their debts. More than 50 percent of all farm operators had off-farm employment in the early 1970s—almost twice the proportion of off-farm employment in the early 1920s.[30] In 1977, 1.9 million farmers needed off-farm incomes to survive. USDA figures show that farmers with the largest and smallest gross sales earned the greatest amounts of off-farm income. The 1.2 million farmers in the middle have farm earnings of under $10,000 per year and off-farm earnings of $6,000 to $10,000 per year, resulting in an average net annual income of about $12,000. Ironically, most of these small-scale farmers own farmland, buildings, and equipment worth over $200,000.[31]

At first glance, it seems that family farmers who have off-farm incomes should have greater financial independence than families who rely solely on farm income, but the jobs they hold in rural areas and small towns are in the secondary labor market and pay low wages. These family farmers are treated like women workers in general: lower wages are "justified" by the assumption that their employment is a secondary source of income.

With so many farmers relying on such large amounts of off-farm income, family farming as a self-supporting way of life is more illusory than real. The buttressing of family farming by off-farm work is an example of how the work ethic itself is mythical. Hard work begets not self-sufficiency and independence, but more hard work and dependency on off-farm employers.

Loss of Decision Making: Contract Farming

The independent management of farms is critical to the concept of the family farm. Marshall Harris, an economist with the USDA and the University of Iowa, said of the independent farmer that "no one tells him what to produce, how much to produce, what inputs to use, or what, where, when, and at what price to buy inputs or to sell products."[32] To protect themselves against widely fluctuating incomes and the inability to get credit from banks, many farmers engage in contract farming. When farmers sign contracts with canning and processing companies, they can retain the illusion of independence because they are self-employed but they effectively become employees of the companies. Under contract farming, farmers provide land and labor and make lower-order decisions while the companies provide everything else—seeds, fertilizers, and usually the harvesting equipment—and make major decisions about planting, spraying, and harvesting schedules. The transfer of farm management decision making to off-farm firms is occurring at different rates across the countryside, depending on the type of farm. Two percent of food grains and 90 percent of poultry are produced by contract farmers. Contract farming is the most obvious example of how family farmers are in fact industrial field workers, stripped of their managerial decision-making powers and left with only their nominal independence.

As all family farmers continue to buy more machines or use more hired labor, rely more on borrowed capital, and lose control over long-term management decisions, they cease to be family farmers and become instead outdoor production workers similar to industrial workers. It would be easy to see the

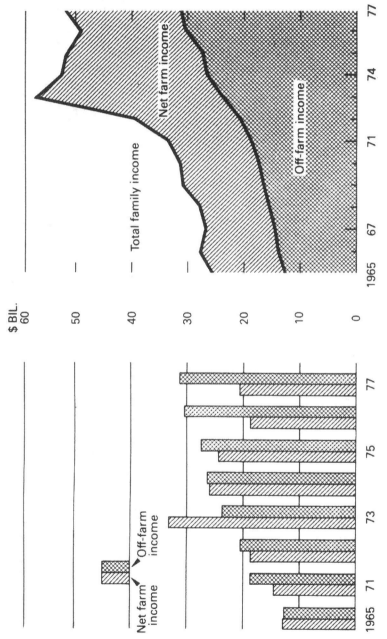

Figure 2.6. Net farm and off-farm income, 1965-1977.

current crisis in agriculture and the demise of family farming in only contemporary terms, as recent phenomena that contradict historical American values. But an examination of the history of land grants in America reveals a pattern of events that is more supportive of family farming as a myth than as a reality.

Part 2

Consequences of Federal
Land and Water Policies:
The Dominance of Agribusiness

3
U.S. Land-Granting Policies

An understanding of current agrarian problems requires a review of major land-granting policies of the federal government in the nineteenth and early twentieth centuries. These include the Preemption acts, the Homestead Act, railroad land grants, the Swamp Lands Act, and the Desert Land Act (see Figure 3.1). The stated objective of these acts was to provide equal opportunities to resident yeomen farm families; in reality, speculators, land companies, and large-scale farmers received most of the land and its associated wealth. A persistent set of proagribusiness institutions and economic interests assured that regardless of the stated philosophy of the land-granting laws, wealthier groups of U.S. society were enriched and the poor were left with dreams.

Before the U.S. government could even distribute its vast land holdings in the interior, it had to devise an easy and quick method of surveying the land. The township and range land survey system was revolutionary in its intent and potential: equal blocks of land were divided to provide equal access to the land. In the early stages of national development, land was the critical form of wealth for new and old immigrants alike. With a widely held and egalitarian land base, the democratic institutions of the new nation were to be guaranteed; economic democracy was to sustain political democracy. This was the revolutionary ideal that inspired people around the world to come to the United States or to copy its ideas elsewhere. Alas, the Jeffersonian ideal was already unattainable by the eighteenth century, because the equal distribution of land would have undermined the dominant interests of industrialists, bankers, and agricultural estate owners.

U.S. Land Survey System

When Europeans conquered North America, some tried to establish feudal landed societies. The French aristocracy in Quebec and the Spanish nobility in California and the Southwest were the most successful in reproducing their feudal order in the New World. To a lesser degree feudal estates were also established in other parts of the United States, particularly in Pennsylvania and in the South. But other settlers preferred, whenever possible, to escape the abusive powers of the landed gentry and the clergy and to settle on scattered

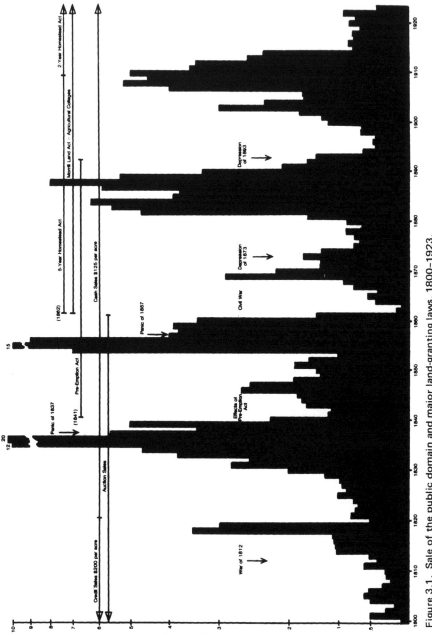

Figure 3.1. Sale of the public domain and major land-granting laws, 1800–1923.

farmsteads. Political and religious leaders attempted to create compact villages in New England like those in Europe; but by the second generation,
settlers had already moved into the countryside to live on individual farmsteads. By the eighteenth century, individual small-scale farming became the
dominant mode of agricultural production along the eastern seaboard. This
tradition became the national ideal as European settlement spread across the
continent.

For Jefferson and other eighteenth-century intellectuals, a nation of small
farmers would provide political freedom, independence, self-reliance, and the
ability to resist political oppression.[1] In their minds, these goals were predicated on the right to own property, especially land. The right to land, the
primary form of wealth in the eighteenth century, meant the right to a job
and economic independence. Jefferson had the foresight to realize that political freedoms would mean little if they were not based on a secure economic
foundation. He reasoned that in a democracy access to land must be provided
by the national government.

> Whenever there are in any country uncultivated lands and unemployed
> poor, it is clear that the laws of property have been so far extended as
> to violate a natural right. The earth is given as a common stock for man
> to labor and live on. If for the encouragement of industry, we allow it
> to be appropriated, we must take care that other employment be pro-
> vided to those excluded from the appropriation. If we do not, the
> fundamental right to labor the earth returns to the unemployed.

The Ordinance of 1785, the first draft of which had been prepared by a
committee chaired by Jefferson, created a procedure for achieving at least
some of Jefferson's ideas. The ordinance authorized the survey of the public
domain (all lands that were at any time owned by the United States and subject to sale) ahead of settlement into six-mile-square townships and square-
mile sections of 640 acres (Figures 3.2 and 3.3). This township and range system
created a new kind of space that was in conflict with the older, conservative
spatial order established along the eastern seaboard. Isonomic space replaced
eunomic space as the ideal. *Eunomy* means "civic order under good laws"; it
signifies an order in which justice was dealt out "according to merit." Under
the previous eunomic metes and bounds survey system, the amount and
quality of land were apportioned according to the merits of the receivers.
Thus eunomic space (the basis for the creation of the landed aristocracy in
Europe) is characterized by irregularly sized and shaped farms. The township
and range system, however, represented isonomic space. *Isonomy* means
"equality before the law" and signifies a social order in which justice produces
an equal division. Isonomic division of the land thus resulted in the equal
distribution of regular-sized and -shaped farms. It was meant to represent the
Jeffersonian ideal of agrarian democracy, and from 1785 to 1860 isonomy
became the overriding influence on the spatial organization of the United
States.

Yet the equal division of the land proved inadequate for the achievement

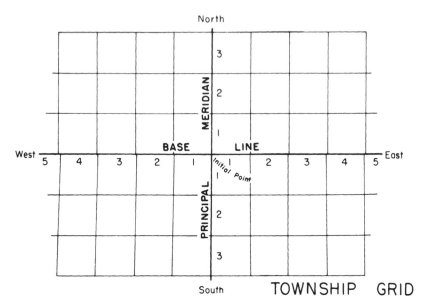

Figure 3.2. The Ordinance of 1785 established the township and range survey system, which is found throughout the United States, except in the original thirteen states and Texas. *Townships* run east and west of the principal meridian; *ranges* are tiers of townships running north and south of the base line.

of economic and political equality. While consideration of land quality in eunomic space had resulted in unequally sized farms and little democratic distribution of land, lack of such consideration in apportionment under the township and range system resulted in equally sized farms of decidedly unequal value. A 160-acre farm in the sand plains of central Wisconsin was hardly comparable to a 160-acre farm in the short-grass prairie of Illinois. Furthermore, the alienation (sale) of the land—the fundamental issue for Jefferson—reflected the landed interests of the day. Despite the Jeffersonian ideal, Alexander Hamilton expected speculators and land companies, not individual farmers, to become the principal buyers of public lands and in turn sell the land to actual settlers at a profit. His expectation became the rule. Not until the mid-nineteenth century were any limitations placed on this wide-open land-granting system.[2]

The Ordinance of 1785 stipulated that one-half of the townships were to be sold as a whole and the other half in sections of 640 acres (see Figure 3.3). Making large tracts of land available, even if sold at auctions for a minimum price of one dollar per acre, assured that speculators and land companies would benefit more than individual settlers. The auction system had the appearance of fairness but actually favored those who purchased large tracts of land for speculation and penalized buyers of small parcels of land for cultivation. Wealthy buyers could outbid and buy more land than low-income settlers because there were no acreage limitations on purchases. Two hundred

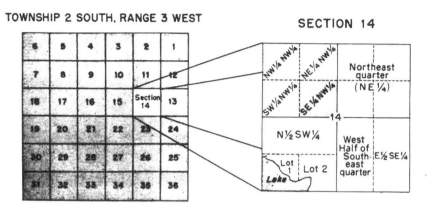

Figure 3.3. Agrarian democracy was based on the equal division of the land. Nineteenth-century land grants were based on the section (640 acres) or parts thereof.

twenty million acres of land were sold for cash and credit by this method (see Figure 3.4). The provisions of the act were not surprising, given that the public domain was to generate revenue for the federal government and that members of Congress, such as George Washington, were large landowners themselves or shared the views of those who were.[3]

The federal government sold these large parcels of land at a much higher price than it had paid for them. Again, the higher prices discouraged and often prevented actual settlers from purchasing public lands. Land acquired by the federal government through the Louisiana, Florida, and Gadsden purchases; the acquisition of the Mexican and Oregon territories; and the annexation of Texas cost in money payments, including interest, four and a quarter cents per acre; yet this same land was being sold under the Ordinance of 1785 to settlers for over one dollar per acre.[4]

A pattern of land speculation, monopoly, and inside dealings soon appeared. Land sales to many large land purchasers and to private companies, which in turn raised the price of land sold to actual settlers, was a common way of disposing of land in the early settlements along the Atlantic and in the colonization of the interior. The Ohio Company was one of the outstanding examples of these private land companies. Congress effectively suspended the Ordinance of 1785 to sell 1 million acres for approximately ten cents an acre to this syndicate.[5] Speculators were also assisted by the first secretary of the treasury, who accelerated the sale of large tracts by making land available at thirty cents an acre, with credit for purchases in excess of ten square miles.[6] By repeatedly ignoring the act, the government provided greater advantages to speculators and land companies than to settlers.

Although Congress had granted credit for large land sales, such as the one to the Ohio Company, it opposed liberal credit for settlers, thus blocking the purchase of land by would-be settlers until the passage of the Preemption and Homestead acts. The interests of inside dealers therefore continually

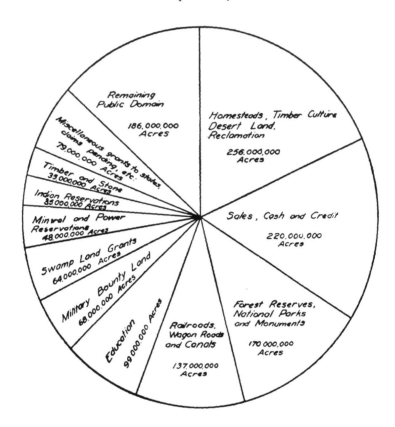

Figure 3.4. Disposition of the public domain: total acreage, 1,399,000,000. *Source:* Benjamin Horace Hibbard, *A History of the Public Land Policies.* Madison: University of Wisconsin Press, 1965, Chart II, p. 570. Reprinted by permission of the publisher.

clashed with actual settlers who could not afford to purchase whole townships or even a whole section as provided by the Ordinance of 1785. Pressure from settlers and some members of Congress gradually reduced the parcel size to 320 acres (in 1800), 160 acres (in 1805), and finally 80 acres in 1820. Even though an 80-acre farm could be acquired for $100 cash, the private sale of a tract of 640 acres went for about $400 cash. Thus, a parcel of land eight times as large cost only four times as much— an advantage for affluent land dealers.

Preemption Acts, 1830–1891

Although the size of tracts had been brought within reach of ordinary settlers, paying for the land was still a problem. The cash system prevented the poor from possessing land. Many settlers, therefore, took up virgin land without paying for it and claimed it by virtue of their labor and improvements.

"Squatters," as these settlers were known, had the sympathy of western politicians and eastern labor organizations who pressed for the passage of the Preemption Act of 1830. It authorized the purchase of up to 160 acres of unsurveyed public land by settlers for $1.25 per acre after fourteen months of residence on the land. In 1832, the sale of 40-acre tracts became available under the act. Both laws acknowledged that the poor had not benefited extensively from the national land-granting policies, which were based on cash sales and large tract purchases. The supposed advantages of these acts were to allow the poor to settle the land and, by working it, produce enough wealth to pay for it.

Northern labor organizations supported the preemption acts because workers could seek a new future on the frontier. Those who took up land in the West helped reduce the number of urban workers, who could therefore demand higher wages. In the West, politicians were sympathetic to squatters because they built farms, roads, schools, and towns—adding people and wealth to the region. In contrast, northern industrialists and southern plantation owners opposed the preemption acts. Industrialists feared lower profits because a smaller urban labor force would demand higher wages and better working conditions, and planters knew that squatting would provide blacks with the chance to escape slavery. Congress was also hostile toward squatters. Congressional voting on the Preemption Act clearly reflected the conflict between the new capital of the West and the old capital of the East. The arguments against the principle of preemption in Congress were (1) the government would lack the authority to enforce law and order; (2) the labor supply of the eastern states would diminish and thus wages would rise and profits fall for industrialists; (3) the principle of public lands as a source of revenue would be destroyed; and (4) preemption would provide a bounty to the new, western states that was denied to the older, already settled eastern states.

Squatters were warned repeatedly to stay out of the public domain, although it was unused and unsettled by Europeans. The federal government went so far as to instruct the militia to move against the squatters and burn their cabins and their farms. Despite these oppressive actions, settling land ahead of surveys continued to be common practice. In 1838, for example, twenty to thirty thousand people had squatted in what is today Iowa, even though no land had yet been offered for sale. Squatters were ready to resist any effort on the part of the government and outside purchasers to conduct auctions at which squatters' interests were not safeguarded.

Despite the supposedly good intentions of the preemption acts, the drive for unearned profits resulted in extensive manipulation of the acts by speculators, "culminating in one of the most outrageous land swindles perpetrated by the federal government."[7] Although the acts were improvements over earlier attempts to auction land, which had largely benefited speculators, they were still open to avoidance and fraud due to speculative pressures.

One way to avoid the acts was to buy Indian lands, which were held in trust by the federal government through executive treaties. Preemption laws protected neither white settlers who occupied Indian lands nor Indian land holdings from these executive treaties. By the time of the Civil War nearly

7 million acres of Indian land had quickly passed into speculators' hands by means of these federal treaties.

Another way of evading the preemption laws was to obtain land under Spanish and Mexican land claims. In the territories that were bought by and annexed to the United States, large French and Spanish private land claims existed. For example, as Mexican authority waned in places like California, Americans were taking over many of the vast Spanish estates. Congress and the Supreme Court took the position "that the right of private property in land in the acquired territories should not be affected by the change in sovereignty regardless of treaty stipulation."[8] A total of 34.6 million acres of private claims was consequently honored and hence excluded from the public domain.[9] In California, Spanish estates, or latifundios, were "protected" from the preemption acts even though valuable valley land was held by only 813 claimants, 87 of whom had obtained their grants during the last six months of Mexican rule.[10] The tracts ranged from 4,000 to 50,000 acres—with a few extraordinary grants as large as 1,750,000 acres—and totaled 14 million acres.[11] Almost a quarter of these claims were rejected for fraud, were antedated (based on forged documents), or were contrary to Mexican law; the rest were honored, allowing last-minute land transfers to avoid the preemption acts.

Some Spanish land grants were subdivided before the end of the Mexican period, especially in northern California where population pressures allowed huge profits to be made from the sale of undeveloped land. But the division of large tracts did not mean the creation of small farms of 100 and 200 acres; instead, bonanza farms of 500 to 5,000 acres were created, particularly in the Sacramento Valley. California, in fact, had been a state for seven years before an acre of land was sold, and of the 8.7 million acres in farms in 1860 only 345,000 were acquired through purchase from the federal government.[12] Land historian Wallace Gates remarked that "the original claims were to have a marked effect on the agricultural character of the State."[13]

Among the California estates that continue to exist today are the Kern County Land Company with 389,000 acres; the Tejon Ranch in the southern San Joaquin Valley with 79,000 acres, now owned by Standard Oil Company of California; the Irvine Ranch with 110,000 acres in Orange County; Rancho California with 97,000 acres owned by Aetna Life Insurance and Kaiser; and the Newhall Ranch with 43,000 acres north of Los Angeles. These and other large farms are associated with major agrarian problems in the West today. James Bryce described these land- and labor-related problems as early as 1889: "The land system of California presents features both peculiar and dangerous, a contrast between great properties, often appearing to conflict with the general wealth, and the sometimes hard pressed small farmers, together with a mass of unsettled labour, thrown without work into the towns at certain times of the year."[14]

Although the U.S. Constitution guarantees civil rights, capital accumulation—profits acquired through land acquisition—reigned supreme. The entrance requirements for Utah and California statehood illustrate this point. Although freedom of religion was protected by the Bill of Rights and the separation of church and state by the Constitution, Utah was forced to deny Mormons their religious right to practice polygamy before being allowed to join the Union. California, on the other hand, was allowed to enter the Union without breaking up its large feudal estates, even though these subverted the goals of the Jeffersonian ideal as well as federal laws such as the preemption acts. Thus the state sided with private property rights for the wealthy, which were not constitutionally protected, and violated constitutionally guaranteed religious freedom to the bulk of the population.

Since improvements on the land in the form of fields cleared and structures built were not specifically required under the preemption acts, speculators, absentee farmers, and land companies could fraudulently benefit from these acts. In 1882 the Commission of the General Land Office declared in favor of the repeal of the preemption laws because they allowed so many frauds. In Kansas, the Dakotas, Colorado, Nebraska, and northern Minnesota fraudulent entries ranged between 70 and 90 percent of all land claims. After the passage of the Homestead Act in 1862, the preemption acts became obsolete, and in 1891 they were finally repealed.

Homestead Act of 1862

The Homestead Act was and is still thought to have been a fundamentally different kind of land-granting law because land that had been considered a source of revenue for the federal government was now a resource to be given free to settlers. Under the law, citizens or persons intending to become citizens could file claims to parcels of unappropriated public land up to 160 acres. (Single women but not married ones could file claims. Historian Grace Fairchild commented that "it always looked to me as if the government was run by men and all the laws were made for them. So women had to take up claims before they got married."[15]) After paying ten dollars and promising that the land was for actual settlement and cultivation, settlers were granted permission to occupy the land. After five years from initial occupancy, settlers received title to the tracts.

Many organizations, such as the Free Soil Party, Free Soil Democrats, and the National Reform Association, supported the principle of homesteading. Their support recognized the continual conflict between speculators and bona fide settlers. Galusha Grow, the foremost champion of homesteading, said that "the struggle between capital and labor is an unequal one at best. It is a struggle between the bones and sinews of men and dollars and cents; and in that struggle, it needs no prophet's ken to foretell the issue. And in that struggle, is it for this Government to stretch forth its arm to aid the strong against the weak?"[16] Likewise, western politicians supported homesteading for philosophical and practical reasons. For example, a Wisconsin paper justified homesteading this way:

We believe that the adoption of the principle which lies at the bottom of this wise and popular measure . . . would constitute one of the most salutary, beneficent, important and glorious reforms that our

Figure 3.5. Settling the land. (Top) The scramble for free land on the Great Plains in the late nineteenth century. USDA photo. (Bottom) Homesteaders in Nebraska, 1887. USDA photo.

Government has ever sanctioned. It would be a practical acknowledge-
ment of the maxim that the soil of the country belongs to the people,
and break away the barriers which prevent the landless and homeless
from making for themselves an estate and a home. . . . Such a good law
would operate with a democratic equality that would furnish one of the
most beautiful and glorious results which flow from the existence of
democratic institutions.[17]

Despite these prosettler sentiments, opposition to the principle of homesteading was intense. In 1860 President Buchanan vetoed the bill because he
feared that it would "introduce the dangerous doctrines of agrarianism and
the pernicious social theories which have proved so disastrous [for the landed
elite] in other countries [e.g., France and Russia]."[18] The South was vehemently opposed to free land because a system of small farms was associated
with free labor—farm families working their own land. Yeoman farmers required no slaves, and the plantation system of large estates was predicated on
slavery. The Homestead Act would prevent the spread of slavery and plantation agriculture into the West, a decided disadvantage from the perspective
of the white plantation owners. Even with the proslavery proponents absent
from Congress in 1862 (because of the Civil War), the final opposition by
northern capitalists was overcome only when evidence from the General Land
Office showed that land had already ceased to be a source of revenue for the
government. They had feared that the loss of such revenue under the Homestead Act would result in taxation of the affluent. Thus, the debates in Congress indicated that despite the sentiments of settlers, the law was enacted
only when southern capitalists were absent and northern capitalists were
certain their interests would not be harmed.

Even with the best intentions, the Homestead Act was passed at such a late
date that most of the agricultural land in the humid East had already been
settled (see Figure 3.6). Most homesteading took place on the semiarid Great
Plains and arid West where 160 acres were inadequate for a family farm. The
default of these farms helped bonanza wheat farms, such as those in the Dakotas, to be formed. Despite the best intentions of the act, the physical environment in which it was applied presented one more obstacle to its real success.

Furthermore, some of its provisions, in combination with a lax and cor-
rupt administration, allowed fraud and speculation to continue. Lawyer Sheldon Greene found that "a federal commission estimated that 40 percent of
the five year homesteads were fraudulently obtained due to faulty administration, false residency, and superfluous improvement and cultivation acceded
to by government officials."[19] The commutation clause, which allowed
homesteaders to postpone meeting the requirements of the Homestead Act
under emergencies, was one source of evasion. Between 1881 to 1904, 23 percent or 22 million acres were assembled into large land holdings through the
commutation clause.[20] Although the purpose of commutation was to assist
homesteaders in times of sickness, crop failure, or inability to make a living
off the land, in actual practice it became a loophole by which speculators
could avoid the purpose of the act. The Land Office reported,

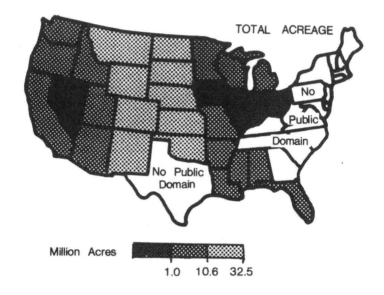

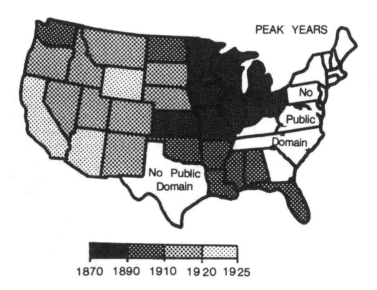

Figure 3.6. (Top) Total acreage homesteaded, 1868–1961. (Bottom) Years of maximum acreage homesteaded. *Source:* U.S. Bureau of Land Management, *Homesteads* (Washington, D.C.: Department of the Interior, 1962).

Actual inspection of hundreds of commuted homesteads shows that not one in a hundred is ever occupied as a home after commutation. They become part of some large timber holding or a parcel of a cattle or sheep ranch. . . . They are usually merchants, professional people, school teachers, clerks, journeymen working at trades, cow punches, or sheep herders. Generally these lands are sold immediately after final proof.[21]

Other kinds of fraud also prevailed. Children were picked up and placed over the number 21 marked in chalk, so that when the government land agent asked them if they were over 21 (a condition of the Homestead Act), they could reply yes. "Houses" of cardboard boxes, 14 by 16 inches, were placed on the land, and oaths were taken that a good board house 14 by 16 (everyone assuming feet) had been erected—another requirement of the Homestead Act.

In the end, the ineffectiveness of the act was assured not so much by its own provisions or administration or even where it was applied, but rather by numerous simultaneous countervailing laws that were subsequently enacted. Gates summarized these conditions:

The retention of the Pre-emption Law and the commutation clause of the Homestead Law made it possible for timber dealers, cattle graziers, mining interests, and speculators to continue to acquire lands through the use of dummy entrymen, false swearing, and, often, the connivance of local land officers. . . . The Desert Land Act, the Timber Culture Act, and the Timber and Stone Act provided even greater opportunities for dummy entrymen to enter lands and assign them to hidden land engrossers. The palpable frauds committed and the large acres transferred under these acts and their interference with the homestead principle lead one to suspect that their enactment and retention were the results of political pressure by interested groups.[22]

During the period of the Homestead Act, Congress passed a series of laws that furthered the interests of speculators directly. Gates cited the major areas that benefited from these laws (see Table 3.1).

1. Over 128 million acres were granted to railroads and another 2 million for wagon roads and canals.

By granting lands to railroads after 1862, Congress from the outset struck a severe blow at the principle of free homesteads. In eight years of the passage of the Homestead Law, five times as much land was granted to railroads as had been given in the twelve preceding years. . . . Such imperial generosity was at the expense of future homesteaders who must purchase the land.[23]

2. Grants to states by the federal government were also contrary to the homestead principle. After 1862, over 140 million acres were in the hands of

TABLE 3.1
Land-Granting Laws that Undermined the Homestead Act of 1862

Land Grants	Land Dispositions After 1862 Million Acres
Railroad Lands	128
Wagon Roads and Canals	2
State Lands	140
Indian Lands	100-125
Federal Lands for Cash Sales	100
Agricultural College Act	8
Total	478-503

Source: Based on Paul Wallace Gates, "The Homestead Law in an Incongruous Land System," in Vernon Carstensen (ed.), The Public Lands (Madison: University of Wisconsin Press, 1968), pp. 315-348.

the states, which sold these lands mostly to land companies, railroads, and speculators.

3. The cash-sale system, continued after 1862 by the federal government, subverted the intention of the Homestead Act by decreasing the amount of land available for homesteading. Over 100 million acres—some of the most fertile land—were sold in the Great Plains and West Coast. This practice allowed large land estates to be formed.

4. Indian lands continued to be unprotected by any uniform land policy. After 1862, between 100 and 125 million acres of Indian land were sold, representing one-half as much as the total acreage that had been entered under the Homestead Act. Initially these lands were sold in large blocks to groups of investors and railroads, and only later were small tracts sold to settlers.

5. The Agricultural College Act of 1862 authorized the sale of public lands to pay for the construction of land-grant colleges. States having no public land available for this purpose received land scrips that could be redeemed for land in the western states. Eight million acres of this land were sold by eastern states through cash sales. "Probably no other scrip or warrant act was used so extensively by speculators to build up large holdings"[24] and to curtail the homestead principle.

California, which today has some of the largest farms and some of the greatest agricultural problems, illustrates the consequences of circumventing the Homestead Act. Land monopolization in the state dates back to the Spanish and Mexican periods when 588 claims amounting to about 10 million acres were confirmed.[25] Between 1862 and 1880, land sales, warrants, and scrip entries in California were made on an enormous scale—totaling over 7 million acres. An additional 8.5 million acres were disposed of by the state. Not only were 16 million acres removed from homesteading, but also huge farms were amassed during this period. For example, between 1868 and 1871 a Mr. Chapman secured 650,000 acres in California and Nevada through fraud, bribery, false swearing, forgery, and other crimes, for which he was charged but never punished. Eventually, he amassed an empire of over 1 million acres. Other prominent land empire builders in California were Henry Miller, Isaac Friedlander, E. N. Miller, and John W. Mitchell. Together, their holdings came to over 1.4 million acres. Another forty-three land speculators acquired almost 1 million acres in California in the 1860s. As Gates pointed out, "Buying in advance of settlement, these men were virtually thwarting the Homestead Law in California."[26]

Although numerous attempts were made at the national level to prohibit the further sale of agricultural land and the granting of land to the railroads in order to practice the full intent of the Homestead Act, Congress paid lip service to the act while circumventing it with numerous contradictory laws. In the end, the Homestead Act proved to be one more slice of pie in the sky for settlers of limited means who lacked capital or technical skills to start farming and make a living on the semiarid western lands.

Railroad Grants, 1850–1943

Railroad land grants allowed land monopolies to acquire huge unearned wealth from the public domain. From 1850 to 1871 the railroads received 130 million acres from the federal government. This is an area as large as all of the New England states plus Pennsylvania and New York. During the same period, nine state governments gave the railroads another 49 million acres, and an additional 40 million acres that were later forfeited for failure to construct lines within the specified time limits.[27] Although the federal government saved $600 million through lower freight rates from land-grant railroads, the railroads netted nearly $500 million from the sale of land granted to them by the federal government.[28] Furthermore, until 1930 all railroads had received $1.4 billion in additional public aid.[29] Clearly railroad subsidies cost the government more than it saved.

Railroad lands were subject to preemption laws, hence they had to respect the property rights of squatters and select alternative land to pay for the construction of railroad lines. Railroad land grants were provided in checkerboard sections along the length of the railroad, usually within 15 to 60 miles of the track. Because of the noncontiguous nature of the sections, the Northern Pacific's land grants extended, in a 120-mile–wide strip, from Nebraska to California. Until 1887, settlement was excluded from the government sections within and adjacent to the land-grant zones. Consequently, unless pioneers chose to buy on the railroad's terms, they had to settle, in the Northern Pacific's case, at least 60 miles from shipping points. Through their grants and the restrictions in the grant zones, "railroads got just about one-tenth of the United States and for years restricted settlement in three-tenths of the United States. The ratio is much higher in the West, where most of the grants lay."[30] Furthermore, settlers who bought directly from the railroads had the greater financial burdens of higher land costs and lower amenity development.

With the Transportation Act of 1940 the federal government agreed to pay full commercial rates to railroads; in return the railroads were to release their claims to land for which they had not met the grant conditions. However, only 8 million acres of the 12 million acres owed were actually returned to the Interior Department. The Southern Pacific, which illustrates the abusive powers of the land-grant railroads, still holds 3.8 million acres. The original conditions of the grant stipulated that all land granted to the Southern Pacific would be sold by it "at a price not exceeding one dollar and twenty-five cents per acre."[31] The government's intention was that both settler and railroad construction would benefit from the sale of the land at a fair price. But keeping the land has been far more profitable for Southern Pacific than selling it. In 1975 the company reaped $25 million in revenue from oil and gas leases, agricultural leases, grazing rights, and timber sales. Railroad grants consistently excluded all mineral rights, which were to remain in the control of the government. Sheldon Greene has argued that the Southern Pacific's continued exploitation of timber and oil resources is a breach of the initial grant conditions and justifies the return of these lands to the public domain.[32]

Based on a thorough analysis of land grants to railroads, Greene concluded that "the lesson of the railroad land grants after more than one hundred years is that government has been incapable of dealing affirmatively and at arm's length with potentially powerful economic interests."[33] Historian Fred Shannon reached a stronger conclusion:

> It [the people's lobby] should demand that after three-quarters of a century of private profit from public gifts, it is now time for the people to take back the property without further recompense, so that in the future the benefits shall be reaped by the people who paid. Any reimbursement to the people made by the land-grant railroads, to the present, has just been a little interest on the original obligation.[34]

Swamp Lands Act of 1850

The federal government gave the states over 64 million acres of supposed swamp lands, the revenue from which was to be spent on drainage and other improvements. Wholesale fraud began with the states' selection of land. "It was found on examination that 75 per cent of the land claimed was not in any sense swampy or subject to serious overflow."[35] Frequently, the most fertile lands, like the Sacramento Valley, were selected regardless of their water conditions. The states disposed of these lands to railroads, land companies, and speculators. In addition, individual land entries were frequently illegal. Henry Miller, for example, acquired dry grassland in the San Joaquin Valley under the terms of the Swamp Lands Act, which gave alleged swamp lands free to individuals who agreed to drain them. The law provided that

BOX 3.1. THE ATLANTIC PACIFIC RAILROAD AND
ITS NEW MEXICO AND ARIZONA LAND COMPANY

No corporation embodies more of the history and the contradictions of the Southwest than the New Mexico and Arizona Land Company, operating out of field offices in Albuquerque. The company's landowner-ship and extensive subsurface mineral rights total more than 1,353,500 acres. Its seven major tracts of land cover more than 2,100 square miles, an area nearly twice the size of Rhode Island and 60 percent larger than the fabled King Ranch of Texas.

The company's control over this acreage goes back to the post-Civil War Congress, whose legislators understood how to establish continental empires. Congress chartered the Atlantic & Pacific Railroad to build a line from Missouri to the port of San Francisco. Having taken the Southwest from Mexico in the War of 1848, and continuing the extermination of the Indians, Congress began granting the railroad free land upon which to build the line. Despite the gift parcels of land, the Atlantic & Pacific managed to go bankrupt a full century before the debacle of the Penn Central. The company that reorganized the defunct A & P, the St. Louis-San Francisco Railroad, fell heir to the A & P's land grants. Congress did not require the return of the land, even though the A & P couldn't finish the line.

The St. Louis-San Francisco Railroad (The Frisco) understood the importance of keeping its land bonanza intact and separate from the company treasury in case the new company should also fail, so it organized a subsidiary in 1890 to manage the land. It was called the New Mexico and Arizona Land Company. Today, The Frisco owns just over 50 percent of the land company stock and holds three of the five board seats. The other directorships belong to F. G. McClintock, board chairman of the First National Bank and Trust Company of Tulsa, Oklahoma's largest bank, and Leonard Spangenberg, chairman of the board of a Massachusetts investment service.

In 1971, the federal government was finishing the construction of Interstate Highway 40 through central Arizona along a route where, a century ago, it subsidized the building of a cross-country railroad. The government needed some land to build four interchanges outside of Winslow, but this time there weren't any Indians to steal it from. All the land was privately owned, so the federal government had to buy it— 190.5 acres for $92,000. The owner was the New Mexico and Arizona Land Company.

Source: Based on James Rowen, "Land Empires," *New Republic,* January 15, 1972, pp. 17-18.

the land had to be underwater and traversable only by boat. Miller loaded a rowboat onto the back end of a wagon and had a team of horses pull him and his dinghy across his desired grassland. Thousands of acres became his through these fraudulent tactics.[36]

Desert Land Act of 1877

The Desert Land Act provided another way for fraudulent entries to be made and large landholdings to be formed. The act provided 640 acres to settlers who would irrigate these acres within three years after filing. The price was twenty-five cents per acre at the time of filing and an additional one dollar per acre to be paid when the land was irrigated.

Abuse of the law occurred immediately. A few furrows were plowed and water was conducted along them for the sole purpose of being able to say that the irrigation requirement was being fulfilled. In Arizona, Wyoming, and Idaho huge numbers of illegal entries were made. The Survey General of Arizona reported in 1887 that speculators had obtained 5,000 acres illegally under this law, and that over half of those filing desert claims, covering 400,000 acres in total, were from outside the state.[37] The most notorious land grant under this act was masterminded by two land speculators, Haggin and Tevis. They paid desperately poor people to enter phony claims for 640 acres each and then, by transferring those claims to themselves, they were able to acquire title to approximately 96,000 acres of California valley land. They managed to do this before anyone else in the state had even heard of the act because they had inside information from Washington on its passage. To preserve their vast landholdings, Haggin and Tevis incorporated under the name of the Kern County Land Company, which in 1967 was acquired by Tenneco, one of the largest agribusinesses in the United States.

Conclusion

The standard references on the federal land-granting laws agree that speculators, land companies, and railroads have benefited more from all of the nineteenth-century land-granting laws than did actual settlers. Fraud and other illegal actions were frequently employed to gain advantages over the settlers, but more often laws were written so that capitalists benefited directly and legally. The history of these laws shows the superior power of landed interests in the United States over the egalitarian rhetoric that justified their passage in the name of the landless poor. This history also demonstrates how exploitation of the myth of equal opportunity has perpetuated the myth of the family farm while actually distributing opportunity and resources unequally.

4
Federal Water Legislation
and Practices

By the early twentieth century, the abuses of previous land-grant policies were well known, and Congress wanted to correct the mistakes of the past by opening up new land for settlers in the West. Irrigation was expected to allow the West to become a new Midwest, characterized by farms that were owned and operated by families. This transformation would be based on massive federal reclamation (irrigation) programs, which were authorized by the Reclamation Act of 1902.

The first director of the federal Reclamation Service explained the purpose of the act as

> not so much to irrigate the land, as it is to make homes. . . . It is not to irrigate the land which now belongs to large corporations, or even to small ones; it is not to make these men wealthy, but it is to bring about a condition whereby that land shall be put into the hands of the small owner, whereby the man with a family can get enough good land to support that family, to become a proud citizen, and to have all the comforts and necessities which rightfully belong to an American citizen.[1]

Theodore Roosevelt was more succinct: "Every [reclamation] dollar is spent to build up the small man of the West and prevent the big man, East or West, from coming in and monopolizing the water and land."[2]

Under the law, the federal government would provide irrigation water, paid for from the national treasury, to farmers in the West. To assure that as many farm families as possible would benefit, the law had two specific restrictions: (1) a 160-acre limitation, or 320 acres if farm owners were married; and (2) a residency requirement. Federally subsidized water could be received only if no more than 160 or 320 acres were owned and if owners lived on or near the land. The desired consequences of the act were to prevent large-scale farms from being subsidized by the federal government throughout the arid West.

The history of federal water legislation and enforcement practices once again reveals the contradiction between the justification of these laws and the actual beneficiaries. The family farm was used to justify the passage of the

Figure 4.1. Irrigation. U.S. Bureau of Reclamation's Columbia River Basin Project, Washington, showing the 1950 Soap Lake Siphon (25-foot diameter pipe). Harold E. Foss, U.S. Department of Interior and USDA photo.

various water-related laws, yet the ways these laws were written and enforced allowed them to be subverted by financiers, land companies, and absentee owners—in short, agrarian capitalists. A brief account of the major events of the federal irrigation laws reveals this pattern of abuse.

Through improper enforcement of the Reclamation Act of 1902, many landowners who held more than 160 or 320 acres were granted federal water but were not required to sell their excess land. Congress corrected this initial abuse by passing the Reclamation Extension Act in 1914. The secretary of the interior was to determine the price of the land in excess of the 160-acre per-person requirement. Although this act was an improvement over the original legislation, it inadvertently encouraged speculation by financiers who could buy land at the secretary's low price and then watch its value skyrocket after the construction of an irrigation project in its vicinity. Abuse was further facilitated by the Bureau of Reclamation's 1916 ruling that the residency provision, which required water users to live within 50 miles of the project, was applicable only to farmers who had made the original water right application. Subsequent beneficiaries of the projects were not required to reside in the local area. In effect, absentee ownership was sanctioned.

In an attempt to curtail the resulting speculation, the Antispeculation Act of 1926 provided that land held in private ownership by a single owner in excess of 160 acres would be appraised by the secretary of the interior on the basis of its value *prior* to the initiation of the irrigation project. Owners who signed contracts with the federal government to sell their "excess" land within ten years at preirrigation prices would in the meantime receive enough water to irrigate—"a time span which allowed for enough farming profit to satisfy all but the greediest."[3] The purpose of this provision was to prevent "undue enrichment" and to stop profiteers from gouging buyers for the "unearned increments" in land prices contributed by the federal projects. Yet big operators could sign the contracts, farm their lands for ten years with subsidized or free water, and then sell the land for windfall profits.

While many landowners managed to profit from loopholes in the law, some large landowners tried to circumvent the requirements of the acts altogether. On their behalf, Senator Downey of California introduced a bill to Congress in 1947 that would have exempted projects in certain western states from the reclamation laws, particularly from the 160-acre limit. But after a month of hearings, Congress responded to the testimony of organized labor, farm, religious, and veteran organizations and let the bill die in committee. Though outright repeal of the 160-acre limitation was not achieved, the tactic of seeking piecemeal erosion of reclamation policy through successive exemptions of small projects yielded results. In 1952 a law applying to the San Luis Valley in Colorado raised the limitation to 480 acres and stated that the change was to furnish no general precedent. In 1954 Congress exempted the Owl Creek, Wyoming, and Santa Maria, California, projects. In each instance reasons special to the particular project were cited as justification for amending the act. But from 1947 to 1958 eight bills were passed that exempted specific projects from the 160-acre limitation.

When the 1944 effort to exempt California's Central Valley project failed, California wanted to take over the federal project. Owing to the high costs of this tactic and to Interior Secretary Ickes' assertion that the federal acreage limitation would still apply if the project were transferred to the state, California modified its tactics by constructing its own project. But the topography of California offered only one route for moving water southward; so the state asked federal permission to use joint facilities—reservoirs, canals, and pumps—without having to observe federal policy. This request was debated in both houses of Congress for several days, and when the actual contract for the transfer was drafted under the auspices of Secretary of the Interior Stewart Udall and Governor Edmund G. (Pat) Brown, obligations to observe federal policy were omitted.[4]

Since the 1950s the opponents of the reclamation acts have also tried to secure their claims through the courts. In various cases, they have argued that the laws were unconstitutional because they discriminated against large landowners; that the 160-acre limitation constituted a taking of property without justification; that the Antispeculation Act of 1926 nullified the residency requirement as the act did not include such a provision; and that the 160-acre limit applied only to projects built solely for irrigation. But in major legal battles, large landowners have lost. In 1958, the U.S. Supreme Court ruled:

> It is reasonable classification to limit the amount of project water available to each individual in order that benefits may be distributed in accordance with the greatest good to the greatest number of individuals. The limitation [160 acres per person] insures that this enormous expenditure will not go in disproportionate share to a few individuals with large land holdings. . . . In short, the excess acreage provision acts as a ceiling, imposed equally on all participants, on the federal subsidy that is being bestowed.[5]

A federal district court in San Diego ruled subsequently that residency is still a requisite, despite its neglect in the 1926 legislation. The courts have also ruled that the law applies to all federal projects that supply irrigation water regardless of their main purpose—whether it is flood control, electricity, or recreation.

But despite the legal victories to uphold the requirements of the reclamation acts, the legislation has failed, largely because it has not been enforced by the Bureau of Reclamation. In 1976 the bureau delivered 27.9 billion acre-feet (9,085 billion gallons) to 18.1 million people, or 31 percent of the western states' population. But of the nearly 9.3 million acres of farmland served by the bureau's 174 projects, at least 850,000 acres are clearly receiving water in violation of the law, according to government records. Because the bureau relies on local water districts to compile enforcement records, the figures are probably understated. In California alone, 61,000 acres—compared to 3.6 million acres eligible for water service—are getting water illegally, including some of the nation's biggest and richest corporations. And in the seventeen western states, another 2.3 million acres owned by 6,014 persons are labeled "excess" and would have to be sold to family farmers at preirrigation prices if the bureau enforced the law.[6]

Geoffrey G. Lanning, who worked on reclamation cases as an Interior Department attorney in the 1960s and 1970s, explained how such violations occur: "The Bureau of Reclamations deliberately violated or avoided the 160-acre limitation, doing so by failure to administer the law at all or, when pressed, by having its captive lawyers write crude loophole provisions that let big landowners ignore the public safeguard."[7] Testifying before a Senate subcommittee in 1978, a farmer from Montana, who had bought 2,000 acres in 1970, said that she had been instructed to "phony up some deeds to put on file with the Bureau, breaking down ownership into 160-acre blocks so we could have water." What is common practice, she testified, is to put "your mother and your dad, your kids and your cat and whoever on these deeds and the copy is filed so the Bureau can be in compliance."[8] Farmers do not even have to phony up their deeds to continue to receive subsidized water. In 1965, the A. Perelli-Minetti Corporation broke up its 1,909-acre holding into twenty-six separate corporations, with a different farmer-stockholder owning each new corporation. Since each of the new corporate entities had less than 160 acres and was technically owned by a single stockholder who did not "own" any other project land, each was eligible for federal water. According to J. Lane Northland, associate solicitor for reclamation and power of the Department of the Interior, "The fact that each of the new corporations has identical officers and boards of directors at this time is unimportant. . . . The identity of officers and directors goes to the question of operation and management and not to ownership."[9]

Technically the 160-acre limitation is enforced, but for all practical purposes it has been evaded by use of various legal shenanigans such as large partnerships and complicated leases—all testified to by former employees of the Interior Department and farm families. These evasions have been possible because the residency requirement—the part of the 1902 act intended to prevent absentee ownership—never has been enforced.

The limited enforcement of the reclamation acts has increased the polarization of rural classes. In 1959, rural sociologist Lynn T. Smith found that for the United States the average percentage of farm "personnel" belonging to the "lower class" was 31 percent; in California the percentage was 56 and in Arizona 71 percent. In the Imperial Valley, where the reclamation law is not enforced, the "lower class," mostly Mexican Americans, represented 87 percent.[10] But a real appreciation of the magnitude and effects of nonenforcement of the reclamation acts requires discussion of the two best-documented examples—the Imperial Valley and the Westlands Water District, both in California.

The Imperial Valley

In 1928 landowners convinced Congress to build the Hoover Dam (at a cost of $177 million) and the $30 million All-American Canal, which diverts water from the Colorado River into the Imperial Valley. Because the funds for the project came from the federal treasury, the reclamation acreage limitation and residency requirements would apply. In 1933 Interior Secretary Ray Lyman Wilbur in the Hoover administration issued a letter that exempted the landowners from the federal restrictions. The letter argued that the reclamation act refers to water *sold* by the government, whereas the Imperial Valley gets its water at no cost. Northcutt Ely, one of Wilbur's assistants, who actually prepared the letter, later became a retainer for the Imperial Valley Irrigation District. Although this letter of exemption was never cleared with Congress, it provided the "legal" means by which the bureau could circumvent the letter and spirit of the reclamation law.

In 1971 U.S. Solicitor General Erwin N. Griswold decided not to appeal a court ruling indicating that the acreage limitation does not apply to the Imperial Valley. He became "convinced that (a) we would not win the case in the court of appeals, and (b) we should not win it."[11] Griswold's determination was based on "the fact that the Imperial Valley was fully developed long before any federal money was spent to build the All-American Canal."[12] The previous canal ran for several miles through Mexico, which "led to a number of problems," whereas the All-American Canal was entirely within the United States—"undoubtedly an advantage for the Valley."[13] Since the original law specified that "no right to *use of water*" (italics mine) was to exceed 160 acres per landowner, the federally subsidized water delivered to landowners in violation of the acreage and residency requirements in the Imperial Valley clearly ignored the letter and intent of the law.

The results of this evasion are clearly evident today. Beneficiaries of the heavily subsidized federal irrigation system are a small group of wealthy growers who hold most of their land illegally. In 1977 fewer than 700 farmers produced nearly $250 million of farm produce in the Imperial Valley, which includes the sixth richest agricultural county in the United States. Yet 72 operators controlled more than half the irrigated land, and absentee landlords owned two-thirds of the land.[14] According to the U.S. Agricultural Stabilization and Conservation Service in Imperial County, 139 individuals and companies farm over 60 percent, or 300,000-plus acres, of the operations with more than 1,000 acres each in the Imperial Irrigation District. Even fewer "farmers" control the land and benefit from the subsidized water when the overlapping and joint ownerships are considered. The Elmore family, for example, owns 3 companies with a total of over 17,500 acres, and conglomerates like Purex, United Brands, Kaiser-Aetna, Dow, and Irvine Ranch are also present. Absentee ownership controls about 70 percent of the 500,000 irrigated acres in the valley[15] (see Figure 4.2).

The concentration of federally irrigated lands in the hands of large and absentee landowners ignores congressional enactments and parallels the distribution of other federal subsidies. The same federal government that spends millions to make the Imperial Valley fertile free-of-charge also paid millions to landowners *not* to grow crops in the early 1970s. Thus, 500 large growers received $12 million annually in farm subsidies, whereas 10,000 landless residents of the valley eked out an existence on welfare payments totaling less than $8 million. Between 1952 and 1964, millions of braceros toiled in the Imperial Valley at wages lower than any others paid in America. Today thousands of Mexican nationals work legally, compliments of the U.S. Labor Department, in the fields. Their presence keeps field wages below two dollars an hour and provides the growers with a lucrative labor subsidy.

One form of windfall profit mentioned earlier is the land appreciation due to federal irrigation. According to government figures, an average acre of bureau-irrigated land in California is seven times more productive than a similar acre in the Dakotas.[16] In 1971, irrigated land in the valley was valued, conservatively, at $700 per acre more than the same land would be worth without water. A landowner with 2,000 acres, for example, would get a $1.4 million bonanza from the federal government, merely because the land was in the right place. The total estimated land appreciation in the valley attributable to federal irrigation exceeds $250 million—most of it concentrated among large-scale, absentee owners.[17]

The Westlands Water District

The Westlands Water District, with its 600,000 acres in California's San

Joaquin valley, lies in the richest agricultural valley of the United States. This is the largest Bureau of Reclamation project. According to bureau estimates, the minimum subsidy per acre in Westlands is $1,600, which equals $960 million for the district.[18] Unlike the Imperial Valley, Westlands has at least nominally complied with the reclamation law. In 1968 the average farm size was 4,640 acres; after a decade of enforcement, the average farm in Westlands was 2,407 acres. But the "excess" land that was "to be sold to family farmers and the landless has ended up in the hands of insiders, speculators, tax shelter syndicators, and other paper farmers."[19] An example of a "family farm" recently created is a company, of which 20 percent is owned by Nissho-Iwai, that leases land from twenty-five persons and farms almost 10,000 acres, though most of the individual holdings are smaller than 160 acres.[20] Other beneficiaries (see Table 4.1) of illegally held irrigated acreages are Southern Pacific Railroad with over 100,000 acres, Standard Oil with over 10,000 acres and another 7,000 leased, and J. G. Boswell Company with 24,000 acres and total holdings of 120,000 acres statewide.[21]

Although technically 217,500 acres of "excess" land are scheduled to be sold by 1985 in Westlands, the nonenforcement of this law will again thwart the intentions of the reclamation acts. Consider the case of Russell Giffen, who is president of the Westlands Water District board. At one time Giffen operated about 105,000 acres and was said to gross more than $30 million annually. Now he must sell off his excess land at preirrigation prices. Giffen is not unhappy about having to sell—he is 72 and in poor health—but he would be very unhappy if he received the true preirrigation price. He would also rather not sell in 160-acre parcels to bona fide resident farmers. It is much easier to deal with big corporations and large real estate wheeler-dealers. Enter John Bonadelle, a Fresno real estate promoter and subdivider, and C. R. Shannon, a rich Visalia cattle rancher. In one Giffen sale of 927 contiguous acres to six Bonadelle creations called Cantua Agricultural Partners I through VI, each partnership was given nominal title to a parcel of 160 acres or less. All of the partnerships had the same general partners, who farmed the 927 acres as a single unit (see Figure 4.3). The Bureau of Reclamation ruled that this flagrant charade conformed with all the legal requirements of the 160-acre law.

A second Giffen sale involved a complicated two-step shuffle. First, 1,752 acres were sold in sub-160–acre chunks to twelve friends, relatives, and employees of Bonadelle. Eleven days later the same land was sold to relatives and associates of Shannon. The first sale was at an average price of $515 an acre—a price approved by the Bureau of Reclamation. Eleven days later the price was up to $695 an acre. The bureau pointed out that it has no control over resale prices.

Giffen's largest sale to date involved 27,198 acres plus interests in 50,860 leased acres. The buyers included Giffen's son and his present business manager. The total price was $32.2 million for land, improvements, and equipment. A price of $530 an acre for the land itself was approved by the bureau. Gerald Gard, Fresno County assessor, estimates that dry farmland in the vicinity of the Giffen sale, or land that will support a crop such as grain

The dinosaur roars -- the dinosaur lies

THE BIG LIE

The dinosaur that is agri-business in the West is roaring and lying in a desperate attempt to establish an oil-like monopoly of our food supply -- a monopoly heavily subsidized by our tax dollars.

On the heels of court decisions and Secretary of Interior Andrus's strong efforts to use the water law against monopoly, large landowners are raising mountains of money for political purchases, propaganda and ads like this one which have or will appear in many newspapers and national magazines.

READ THIS AD CAREFULLY

**NOTE THE LIES
SHOW THEM TO YOUR FRIENDS
AND NEIGHBORS**

We must, **WE CAN**, stop this big lie attempt to control us through our stomachs.

The big operators have scared, bought off, pressured and/or brainwashed many people including some small farmers like the one in this ad.

READ THIS AD, AND THEN
GET MOVING.

A.

6

B.

A. Why are these folks so upset?

A family of 6 × 160 = 960 ACRES OWNED = 1,920 LIMIT
" LEASED (3 sq. miles)

Why are these folks so "sad"? Under the proposed Andrus rules they would be eligible for up to 1920 acres' worth of subsidized water on farmland "where the sun spends the winter." Even under NLP's strict proposals they could get up to 640 acres' worth of subsidized water. Could it be that these folks are fronting for SP, Irvine, Safeway and the other big corporations?

Lie 1

THE TRUTH: U.S. Federal Court of Appeals ruled the reclamation law applies to Imperial Valley after **THREE YEARS** of study.

Lie 2

THE TRUTH: Under new rules proposed by Secretary of Interior, this family could permanently obtain subsidized irrigation water for up to 960 acres they own (160 for each family member) and for up to an additional 960 acres they rented -- a grand total of 1920 acres' worth of subsidized water. On top of this, they could get subsidized water for up to five years on an unlimited amount of land they might either own or lease.

Lie 3

THE TRUTH: If this family ever sold land as part of a contract to receive the federally-subsidized water, family members would have preference. If no family members were eligible or interested, the land would be sold through a public lottery in which all of us would have an equal chance to share in the public wealth we all helped to create.

The price of the land would be current market value less the project benefits -- in other words, what the land would be worth today without the subsidized water.

Lie 4

THE TRUTH: The water supply was silt-loaded and unreliable -- alternating between floods and low flows because there were no water storage dams on the Colorado River. The heavy silt required huge expenditures to clean out distribution systems. The international canal was subject to appropriation by Mexico.

For these reasons, Imperial landowners actively lobbied for the Boulder Canyon Project including Hoover Dam, three other dams which store water, prevent floods and precipitate silt; one diversion dam and the All-American canal. When the project was finished the Imperial Irrigation District announced:

"It now may be said with confidence, that no section of our nation is more assured of a permanent and prosperous future than is this valley."

Lie 5

THE TRUTH: Imperial Valley landowners do not pay any of the costs for Hoover Dam or any of the other three Colorado dams which provide flood control, water storage and silt precipitation. Most of the dam costs are borne by Los Angeles taxpayers. Imperial landowners are repaying some of the capitol costs of the diversion dam, Imperial, and the All-American canal; however, they are paying no interest on the 40-60-year "loan." Also, the canal began delivering water in 1940. Repayment did not begin until 1955. How would you like to buy a house, live in it 15 years before making any payments and never pay any interest on the mortgage?

B.

Under court order, obtained by National Land for People, President Carter is now considering strict enforcement of this long-ignored law. We can help the law and ourselves.

Write Carter. Ask him to ban absentee owners, limit water to 640 acres' worth and distribute big holdings through a lottery at fair prices.

for more information write
National Land for People
2348 N. Cornelia
Fresno, CA 93711
(209) 237-6516

Lie 6

THE TRUTH: The Boulder Canyon Act of 1928 authorizing all the dams and the All-American Canal said: "This act shall be deemed a supplement to the reclamation law, which said reclamation law shall govern the construction, operation and management of the works therein authorized, except as otherwise provided."

No "otherwise" is provided by any act of Congress.

Nevertheless, Imperial was "granted" an "exemption" from reclamation law during the last two weeks of President Hoover's Administration in February 1933. At the request of the Imperial Irrigation District, an assistant secretary of Interior named Northcutt Ely -- without seeking the legal counsel of the Interior lawyers -- wrote an "exemption" letter. Secretary of Interior Wilbur signed the letter. Two weeks later Roosevelt replaced Hoover in the White House, and letter writer Northcutt Ely began a lifetime career representing the Imperial Irrigation District.

In 1963, Interior Solicitor Frank Barry ruled that Ely's letter was "clearly wrong," thus beginning the long court wrangle which ended in August when the Appeals court ruled that the law applies in Imperial.

Figure 4.2. The big lie about the reclamation act. Reprinted by permission of the National Land for People, 2348 N. Cornelia, Fresno, California 93711.

TABLE 4.1
Major Farmland Owners and Water Users in the Westlands Water District, 1977

Owners and Leasers		Acres
Southern Pacific Railroad (all land except 2,000 acres leased to other farms--most listed below)		106,00(
Standard Oil (all land leased to other farms--most listed below)		10,47(
	Sub-total	116,47(
Boswell (Boston Ranch, owned entirely by J. G. Boswell)		26,48!
Telles		24,44(
Harris Farms (world's largest cattle feed lot)		18,39:
Producers Cotton Oil (Bangor Punta)		14,78(
Diener Family		11,06:
Westfarms		10,52:
Airway Farms		9,70
Coit		9,14:
Giffen, Price		8,96(
Murrietta Farms		8,65!
Gragnani		8,43(
McFarland (Vasto Valle)		8,30(
Jubil Farms (Nissho-Iwai American Corp.)		7,53(
J. Woolf (former Giffen manager)		7,18(
Sumner Peck (Giffen in-law)		6,59!
O'Neil, Jr. (WWD Board)		5,95!
Jack Stone (WWD Board President)		5,88!
Dr. R. Burford		5,85:
McCarthy		5,71!
Coelho		5,70
Dudley Frank Farms		5,49!
Allen Farms (WWD Board)		5,45(
J. Lowe (past Giffen employee)		5,01
	Sub-total	225,29(

Source: National Land for People, information sheet, based on water billing records from the Westlands Water District in October 1979.

without much irrigation water, has a market value of $250 to $300 an acre. I: Gard is correct, the bureau's calculation of Giffen's "prewater" price is abou 100 percent too generous. In all of the above sales, none of the purchaser: was a bona fide resident of the land acquired. Despite court rulings upholdin₡ the residency requirements, the bureau neglected—as it always has—this im portant section of the law.[22]

The purpose of the original reclamation act and the subsequent relatec acts was to distribute the benefits of landownership and water availabilit₁ as widely as possible. The reclamation program was intended to enrich th₍ lives of working farm families, not simply the owners of land, and certainl₁ not large-scale, absentee owners. Former Secretary of the Interior Stewar Udall pinpointed the problem with the reclamation program when he saic before a Senate subcommittee, "The fact of the matter is, as I look back or my service in Washington twenty years ago, we paid lip service to the famil₁ farm but we really worshiped at the altar of agribusiness."[23]

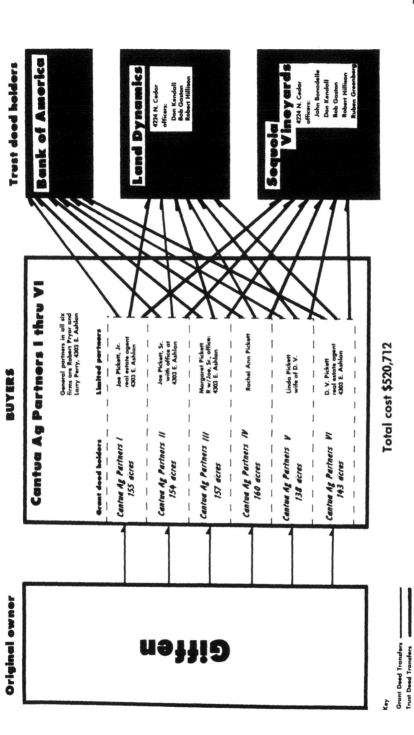

Figure 4.3. An example of the questionable land transactions in the Westlands Water District, California. Reprinted by permission of the National Land for People, 2348 N. Cornelia, Fresno, California 93711.

Yet the friends of agribusiness are still helping to undermine the reclama
tion acts. For example, California's conservative Senator Sam Hayakawa ha
introduced into Congress the "Family Farm Liberation Act of 1979." Whei
questioned by reporters, Hayakawa admitted that if his bill passed, the liber
ated parties would not be family farmers. The bill would release big western
land companies from the reclamation law by removing the residency require
ment, allowing unlimited leasing, placing no limits on the size of farms tha
would receive federally subsidized water, and allowing big land companies t(
own directly unlimited irrigated acres if they pay "full cost" for the water
Although the actual full cost is twenty to thirty times the subsidized price
Hayakawa predictably says that full cost is only about twice the current sub
sidized prices.[24]

Paul S. Taylor, a longtime analyst of federal reclamation projects, con
cluded that "the central problem arose from the fact that potentially irri
gatable lands had largely passed into private hands long before public reclama
tion became a reality. This created special interests resistant to the control:
over monopoly and speculation incorporated in the 1902 reclamation law."[25]
In addition, no civil or criminal penalties were ever prescribed for violation oi
the law by a recipient of project water.[26]

By 1970 the Bureau of Reclamation had spent almost $10 billion to irrigate
nearly 7 million acres. Despite the provisions of the reclamation act, lanc
monopolies are firmly entrenched in the West. The bureau and the court:
have repeatedly refused to enforce the act, with the consequence that huge
absentee-owned corporate estates receive subsidized water while small resident
farmers—the intended beneficiaries—are all but absent in the West.

This Land Is Not Our Land: Who Owns Rural America?

Jeffersonian agrarian democracy was based on the even distribution of agricultural land and was intended to go a long way toward equalizing wealth and spreading decisions widely among people. Yet this goal has rarely reached fruition; the ideal of the family farm has frequently been ignored and land empires have been created instead. For example, nineteenth-century land-granting policies and practices created large land holdings. Today railroad, timber, and energy companies are major land resource owners. Four railroads own 23 million acres in surface and mineral rights. The twelve largest land-holding timber companies own 35 million acres. U.S. oil corporations own 65 million acres, excluding their offshore acreage. Midland Coal, a division of American Smelting and Refining Company, owns approximately 55,000 acres of land in central Illinois. Through its landholding and agricultural subsidiary, Meadowlark Fawns, AMAX Coal Company, a part of American Metal Climax, controls more than 100,000 acres of land in Illinois and Indiana. In 1977 this land produced 13 million bushels of grain and 1.4 million tons of meat![1] Table 5.1 reveals only the tip of the iceberg of corporate ownership and control (through leases and options) of U.S. land resources. Only two federal agencies, the Bureau of Land Management and the Forest Service, control more land than Exxon. Standard Oil of Indiana owns more land than the National Park Service, and Champion has more land than the U.S. Army.

Rural land is a vital resource, even in an urban, industrial society. Private and public owners of land influence the prices we pay for food; they control the amount and kind of open space for private recreational use; and they control timber supplies. Land resources themselves provide basic societal needs—food, timber, fuels, minerals, and amenities—and hence, owners of land acquire income (in the short term), wealth (in the long term), security, and status. Land is a means of distributing and exercising power. The vital importance of land inevitably and necessarily makes the question of who owns the land critical.

About 60 percent (1.3 billion acres) of the land area of the United States is owned by private individuals or corporations and the remainder (or 1 billion acres) is owned by local, state, and federal governments. Each type of owner controls different kinds of land. Virtually all agricultural land, repre-

TABLE 5.1
Corporations with Major Land Holdings in the United States

Oil and Gas Holdings		Acres (in millions)
Exxon		40.2
Standard Oil of Indiana		27.5
Gulf Oil		12.5
Shell Oil		9.5
Standard Oil of California		9.0
General Crude Oil		4.8
Diamond Shamrock		1.7
Amax		1.2
International Paper		.5
	Sub-total	106.9
Timberland		
Champion International		17.7
International Paper		7.7
Boise-Cascade		6.0
Weyerhauser		5.7
St. Regis Paper		3.5
Georgia Pacific		3.4
Crown Zellerbach		2.1
Union Camp		1.7
Continental Forest Industries		1.5
Diamond International		1.4
ITT Rayonier		1.3
Time		1.0
Packaging Corp. of America		.5
Forest Product		.2
	Sub-total	53.7
Mineral, Timber, and Agricultural Holdings		
Burlington Northern		8.6
Union Pacific		7.9
Southern Pacific		5.2
Tenneco West		1.1
	Sub-total	22.8
	Grand Total	183.4

Source: Peter Meyer, "Land Rush," Harper's, January 1979, p.
47; and Fredrick J. Parella, Jr., Poverty in American Democracy
A Study in Social Power (U.S. Catholic Conference, 1975).

senting 20 percent of all land, is held by private owners, the result of nineteenth-century land sales and land grants by federal and state governments. One-third of U.S. grazing land is in the public domain, but actually most of this land (more than a quarter of all land) is treated by ranchers and federal agencies as an extension of privately held lands, which control water, winter range, and access. For agricultural (farm and ranch) purposes, virtually all grazing land is in private hands (see Figure 5.1).

Gene Wunderlich, economist with the U.S. Department of Agriculture (USDA), has studied landownership patterns in the United States for over a decade, using whatever data have been available. He noted that of the 1.3 billion acres in private hands in the United States, only 2 percent, or 26.3 million acres, is residential. The 26.3 million acres, having the broadest distribution among individual landowners, are owned by some 50 million "entities." From the available data, Wunderlich had to rely on information about parcels of land rather than on people. Thus, he had no way of knowing whether numerous entities/owners are not in fact the same individual or, more likely, the same corporation. Another 3 percent, or 40 million acres, of private land is commercial, industrial, and wasteland and is owned by about 3 million entities. Finally, 95 percent, or 1.2 billion acres, is privately held farmland, ranchland, and forestland, but is owned by only 7.5 million entities. Assuming that each entity is indeed a different individual, a generous interpretation of Wunderlich's data would indicate that 3 percent of the population owns 55 percent of the total U.S. land area and 95 percent of the private land![2]

The Economic Research Service of the USDA found that in 1978, 568 companies controlled, either directly through ownership of title or lease or indirectly through the purchase of mineral rights, 302 million acres in the United States. This represents more than 11 percent of the total land area of the entire country and 23 percent of all privately held land. On a global basis these companies controlled almost 2 billion acres, an area larger than Europe.[3]

Landownership in Selected States

Who owns rural America? This question cannot be answered precisely. Accurate data do not exist even though the U.S. government collects enormous quantities of information. The few specific studies that have been made of landownership indicate the general direction of ownership patterns in the United States. They also demonstrate the significance of acreage as a farm classification variable. Examples are selected from states with traditionally land-concentrated patterns and from states that most people think have predominantly small landholdings. These regional studies confirm that land is disproportionately concentrated in the hands of nonfarm corporations, absentee landlords, and large-scale producers. The concentration is occurring not only in states with traditions of large landed estates (California, Texas, and Florida), but in eastern and midwestern states as well.

California

In 1973, twenty-five landowners held more than 61 percent (or more than

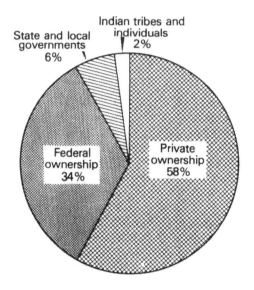

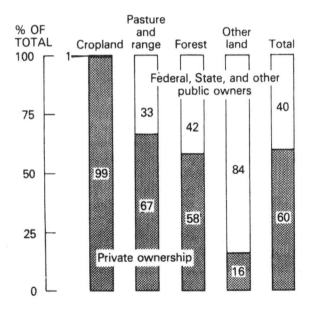

Figure 5.1. U.S. landownership (top) and major land uses by ownership (bottom), 1974.

8 million acres) of the state's private lands. On a county-by-county basis, Fellmeth found that the top twenty landowners in rural counties, constituting a fraction of 1 percent of the population, generally own 25 to 50 percent of the land.[4] Nearly 4 million acres of farmland are owned by forty-five corporations; one corporate farm, Tenneco, controls more than 1 million acres alone. Among other large owners are Standard Oil, Southern Pacific, Times Mirror Corporation, Penn Central, Boise-Cascade, and Leslie Salt.[5]

Texas

Eleven landowners control 5.8 million acres of the state, an area approximately the size of New Hampshire.[6]

Florida

In 1960 only 1 percent of the state's citrus lands were held by large farming-canning corporations. By 1972, 20 percent of those lands were under such ownership.[7]

Maine

Fifty-two percent of the state is owned by about twelve corporations. By one estimate, 80 percent of Maine is held by absentee owners, including corporations with their headquarters outside the state and individuals with their principal residence beyond Maine.[8]

New York

A research group in 1970 found that more than 50 percent of the private land within an upstate study area—one-fifth the size of the entire state—was held by 1 percent of the landowners. Three timber companies owned more than 100,000 acres each.[9]

Nebraska

The Center for Rural Affairs, a private family-farm advocate organization, found that from 1968 to 1973 landownership changes in thirteen northwestern counties, including the Sandhills region, showed that corporate ownership had increased 64 percent. Urban-based farm management companies also grew sharply. The largest farm management company was handling 61,267 acres or 45 percent of all land managed by firms; the top four firms had nearly 72 percent of the business (measured in acres). The study also found that 40 percent of the corporations owning land in the study area were not authorized to do business in Nebraska. By failing to obtain authorization, they escape paying the state occupation tax and avoid record keeping that would publicly expose such corporate dominance of the land.[10]

The Sandhills of Nebraska, which have escaped the center-pivot irrigation boom found over much of the state, today remain completely in rangeland. Despite the Homestead Act of 1862 and the Forest Lieu Act of 1897, land-ownership from the beginning of European settlement was highly concentrated among the wealthy local as well as absentee investors.[11] This pattern of landownership is intensifying today. Nearly 47 percent of the almost half million acres of absentee-owned ranchland was purchased between 1970 and 1976 (see Table 5.2), and all of these recent sales came from previously locally owned and operated ranches. As a result, in Grant County, for example, seven of the sixty-five ranches own 40 percent of the land.[12]

Kansas

A survey of county platbooks for each of the 105 counties in the state indicated that 20 corporations and 164 individuals, partnerships, and tenants owned more than 5,000 acres each. Clark County illustrates the spatial impact of large landholding units (see Figure 5.2).[13] This pattern prevails despite a Kansas law banning corporate involvement in most forms of farm production and placing a 5,000-acre limit on corporate landholdings (see Figure 5.3).

Summary of Landownership Patterns

In summary, wherever landownership patterns are examined in the United States, the nineteenth-century trend of concentrating land in the hands of wealthy absentee individuals and corporations is continuing. Farms exceeding 1,000 acres increased from 28 percent of acres in farms in 1930 to 54 percent in 1969. The increase in large-sized farms was occurring not only in the West, but also throughout the rest of the United States (see Table 5.3). These farms also increased their percentage of total U.S. cropland from 24 to 29 percent and of irrigated land from 44 to 48 percent; yet in 1974 when they held 58 percent of all land in farms, they represented only 7 percent of America's farms. Landownership was even more concentrated in the largest farms: those with 2,000 acres and over represented 3 percent and controlled 46 percent of all land in farms.

The concentration of farmland ownership is associated invariably with an increase in urban-based ownership. Nonfarmers accounted for 33 percent of all farm purchases from 1959 to 1967 and for 38 percent from 1968 to 1976. According to the 1974 Census of Agriculture, 330 million acres of land in farms—almost 40 percent of all private farmland—were owned by nonfarmers.

Land-Tenure Categories

As the number of farms dropped from 6.8 million in 1935 to 2.4 million in 1974, fewer and fewer persons owned more and more of the land and the number of farm families who owned all of the land they worked decreased by 62 percent. Rent payments to nonfarm landlords increased threefold from $491 million in 1940 to $1.5 billion in 1970.[14] Traditional land-tenure categories, based on the Census of Agriculture, group landownership into owners, part owners, and tenants. Full owners—usually mortgaged—own all the land they work; part owners own most of their acreage and lease additional acreage; and tenants own little land and lease most of their acreage. In the past, rural sociologists, land economists, and government officials argued that full ownership was the most desirable and part ownership the next desirable. They reasoned that debt-free, full ownership of farmland allowed maximum decision-making freedom to farm families. But as the price of farmland

TABLE 5.2
The Top Ten Absentee or Part-Time Ranch Owners in the Sandhills of Nebraska

Owners and Acreage	Descriptive Details
Rush Creek Land and Livestock Company: 108,736 acres	Founded in the 1880s by Thomas E. Wells, an early director of the Quaker Oats Company. Rush Creek has grown from 10,000 acres to 155,864 acres in Nebraska. It is owned by Wells' heirs, Thomas Wells, Jr., and Preston A. Wells of Fort Lauderdale, Florida.
Brown Land Company: 45,481 acres	Owned by the Brown family, founders of Brown University, Providence, Rhode Island.
Kiewit Company: 42,949 acres*	Peter Kiewit of Omaha is president of this company. His Peter Kiewit Sons Company is one of the largest construction companies in the nation. Peter Kiewit Sons, Inc., the parent company, has been ranked the tenth-largest coal producer in the U.S. Kiewit also owns the Omaha World Herald.
Harland Milligan: 37,341 acres*	Milligan owns Cornbelt Elevator Company and farmland near Hooper and Scribner, Nebraska.
Big Creek Land and Cattle Company: 33,150 acres*	The owner, J. W. Vieregg, lives in Grand Island, where his wife, Virginia, serves on the school board. Vieregg lists himself as a rancher in Polk's City Directory and as an owner-operator of his Corporation Farm Report. The ranch is reportedly run by a manager.
C. E. Nicholas: 29,162 acres*	Nicholas is a Dillon, Montana banker and rancher.
S and W Cattle Company: 23,177 acres*	Sheldon Wert, a Minneapolis real estate developer and investor, is the owner. He also owns farmland in North Dakota and Minnesota.
Columbian Hog and Cattle Powder Company: 22,979 acres	This Kansas City firm manufactures cattle feed supplements.
Curtis Chisum: 22,505 acres*	A rancher from Dalhart, Texas, Chisum reportedly owns land in Colorado as well as Nebraska.
A. W. Moursund: 21,492 acres*	A lawyer, banker, and rancher from Round Mountain, Texas. A. W. Moursund was the trustee of Lyndon B. Johnson's $14 million estate when LBJ was president.

*Land which includes some of the 200,000 acres purchased from local ranchers between 1970 and 1976.

Source: Center for Rural Affairs, New Land Review, Fall 1976, p. 12.

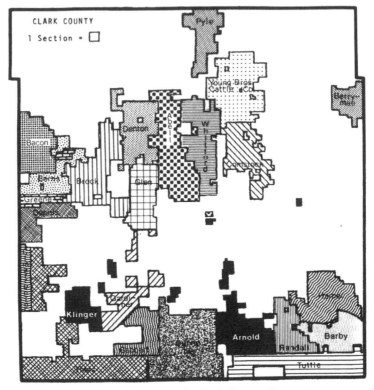

Figure 5.2. Landownership patterns in Clark County, Kansas. Four corporations and twenty individuals or families own 235,000 acres or 38 percent of the county.

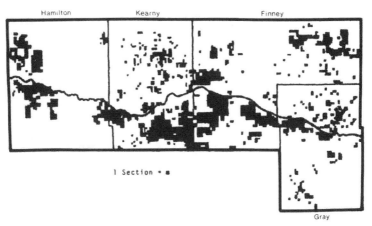

Figure 5.3. Land held by owners of more than 5,000 acres in the Arkansas River Valley, Kansas.

TABLE 5.3
Farms with more than 1,000 Acres as a Percentage of All Acreage
in Farms, 1930-1974

	Percentage of All Acreage in Farms			
Region	1930	1950	1969	1974
Northeast	2.5	3.3	5.5	8.5
North Central	12.3	20.6	33.0	37.4
South	24.9	36.1	47.4	52.2
West	64.8	82.3	89.3	90.0
U.S.	27.9	42.6	54.4	58.8

Source: 1974 Census of Agriculture. Vol. 2, Part 2, p. 11-12,
Table 16.

escalates, the limits of indebtedness are reached, and competition with agribusiness requires still further expansion, more and more farmers are unable
to attain or remain within the full-owner category. Interestingly, academics
and government officials today no longer suggest that full ownership is a virtue
yet still contend that family farmers can remain viable as part owners. This
change of opinion by researchers and policymakers reveals that "facts" are
never so important as the values and analytical frameworks from which the
facts are interpreted. If the myth of the family farm is used to obscure real
trends in landownership, the fact of declining full ownership can simply be
redefined as insignificant. But if family farming is to have any meaning in
its traditional sense, this same fact documents the death of an ideal.

Since 1945, part owners have increased their share of farmland. While full-
owner control dropped from 40 percent of the acreage in 1945 to 35 percent
in 1974, and tenant-operated land decreased from 24 to 12 percent, part
owners increased their share from 36 to 53 percent. Yet that same 53 percent
of farmland represented only 27 percent of all farms. Full owners, in contrast,
who accounted for 62 percent of all farms, controlled only 35 percent of the
land.[15] Thus, although full owners—the true family farmers—appear to dominate the number of farms, they actually control a smaller proportion of land
than do part owners and even less than their own numbers would suggest (see
Figure 5.4).

The above acreage figures indicate that full-owner family farms are most
often small farms and therefore less competitive and less viable. This observation is supported by the data on gross sales of such operations. From 1964 to
1974, the percentage of full owners in the over-$40,000 sales classes increased
only slightly, but the percentage in the less-successful, under-$40,000 classes
increased significantly. At the same time, the proportion of part owners

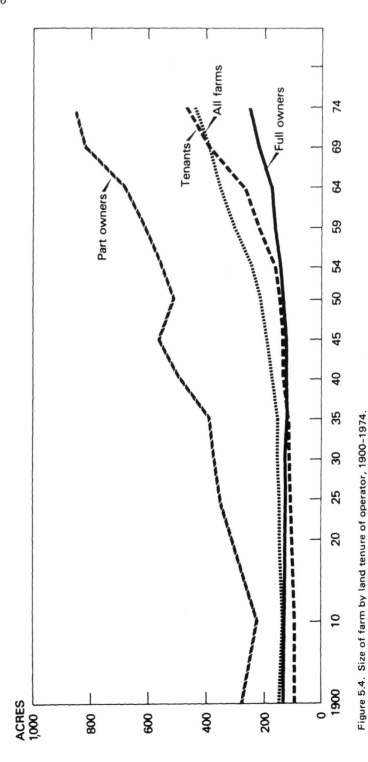

Figure 5.4. Size of farm by land tenure of operator, 1900–1974.

decreased in the lower sales categories and sharply increased in the more lucrative $40,000-plus farms.[16] Apparently, farmers are increasing their sales by renting more land, forfeiting their full-owner status, and therefore maintaining less control over their operations. The full-owner farm family not only controls less of the land, but also reaps less of the reward.

According to the USDA, both full owners and part owners are family farmers. Together they worked over 791 million acres in 1974, 87 percent of all land in farms. But, as Table 5.4 shows, much of that acreage was leased, either from other owners or to other operators. If the definition of family farming includes ownership of the land worked and even if we include tenant-operated land that is owned (which the USDA does not), then only 579 million acres, or 64 percent of land in farms, are truly family farmed. This is a 23 percent decrease in the USDA's figures and does not even consider that much of the remaining acreage is operated by hired managers. Table 5.4 also reveals the conflicting roles of farmers. Many farmers are at once owner-operators, landlords, and tenants; and therefore even those who are basically owner-operators have reason to protect the often opposing interests of tenants or landlords.

Foreign Ownership of Farmland

Since the 1973 oil crisis, farm groups and members of Congress have become concerned about the number of foreigners buying U.S. farmland. Since no comprehensive data is available on the nationality of landowners in the United States, the actual amount and location of all foreign-owned land is unknown. A few studies, however, do suggest the magnitude of foreign ownership of rural land. The General Accounting Office estimated that foreigners invested $800 million to $1 billion in farmland in 1977—about 30 percent of all foreign investment in the United States that year. The USDA estimated that up to 21 percent of all farmland sales in 1977 were made to foreigners.[17] The Senate Agricultural Committee's survey of farmland purchases across the United States during one eighteen-month period found that foreigners bought a total of 826,543 acres—an area larger than Rhode Island. A Department of Commerce study of direct alien involvement in the United States at the end of 1974 showed that 4.9 million acres of land were foreign-owned by 6,000 firms and individuals and that almost 63 million acres were leased by foreign-owned American enterprises. Of that total about 23 percent, or 1 million acres, was identified as farm related.[18]

Many Americans object to foreign ownership of U.S. farmland in principle. Others object to the high land prices that reportedly result when foreigners (those with oil money, for example) are able and willing to pay well above the current market price. In order to monitor the actual occurrence of foreign sales, Congress passed the Agricultural Foreign Investment Disclosure Act of 1978. The act requires that all foreign ownership or transfer of farmland be reported to the federal government and applies to aliens who own as little as a 5 percent interest. A heavy fine of up to 25 percent of the property's fair market value will be imposed for failure to report such ownership.

TABLE 5.4
Acreage Owned and Rented by Tenure of Operators, 1974

Tenure of Operators	Acres Owned and Rented				Rented to Others	
	Owner-Operated Land		Rented from Others			
	Acres	Percentage[a]	Acres	Percentage[a]	Acres	Percentage[b]
Full Owners	302,630,900	99.7	668,277	0.3	25,125,202	8
Part Owners	273,652,477	52	250,789,497	48	11,565,077	2
Tenants	2,622,983	2	116,232,528	98	4,266,276	4
	578,906,360		367,690,302		40,956,555	

[a]Percentage of all owner-operated land and rented land from others, which together equals 100.
[b]Land rented to others as a percentage of all owner-operated land and land rented from others.

Source: 1974 Census of Agriculture, Vol. 2, Part 3, p. I-54, Table 41.

In 1978 the Missouri legislature passed a law that, according to press releases, prohibits nonresident foreigners from buying farmland. On closer examination, Senate Bill 685 actually allows nonresident foreigners to purchase unlimited amounts of farmland and allows them to act as landlords by leasing newly purchased farmland. As R. J. Zani, a resident of Springfield, Missouri, said, "This 'new' law is no more than an artifice to appease (verbally) the xenophobia and racism of rural Missouri, while in fact protecting and maintaining the status quo concerning land sales and speculation in Missouri. It is a clever bit of political pinchbeck, and nothing more. One can rest assured that no profits will be lost for local real estate brokers and bankers."[19]

Much of the debate on foreign ownership of land ignores the critical issues of absentee ownership. While almost 40 percent of all private farmland was owned by nonfarmers in 1974, a more recent General Accounting Office study concluded that "even if foreign owned land is doubled to account for those not tabulated, foreign ownership would amount to less than 1 percent of the 1.3 billion acres of privately held land."[20] George Rucker, research director of Rural America, expressed the issue well, "To the small farmer who is forced to pay rent on his farmland because he cannot afford the escalated real estate value of the land, it matters very little whether the owner of the land resides in Saudi Arabia or on Madison Avenue."[21] Absentee ownership, either by U.S. citizens and corporations or by foreigners, undermines the well-being of rural communities. The fundamental concern should be whether farmland is owned by farm families who live in the local county and work the land themselves.

Disappearance of Black Landownership

Although a great deal of fuss has been made in the national media about the increase in foreign ownership of land, almost no attention has been given to the particularly rapid decrease of farmland ownership by black Americans. The erosion of black landownership has been especially acute because blacks have relatively low rural economic status. The history of black farmland loss is the history of rural neglect and injustice at the hands of agribusiness and state and federal governments.

After Abraham Lincoln's Emancipation Proclamation of 1862, blacks quickly acquired title to land, despite white planters who used the sharecropper system to further enslave blacks. By 1909, 3 million blacks were engaged in farming, and by 1910 black landownership peaked at 15 million acres (see Figure 5.5), and black farmers accounted for almost 70 percent of the total black population and 30 percent of the total southern rural population in 1910.

But the rapid pace of farm mechanization, particularly in cotton, after 1914 facilitated and indeed necessitated massive black migration to cities. As white landowners invested in machinery and large-scale operations, they displaced the sharecroppers whose labor was no longer needed. Small and poor landowning farmers were often unable to compete with the profitability of mechanization. Although this move toward agribusiness affected all small-

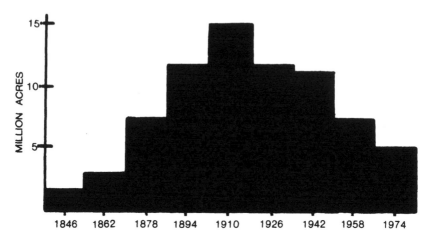

Figure 5.5. Black farm acreage changes, 1846–1974. *Source: Census of Agriculture.*

scale farmers, a disproportionate number of blacks were forced out of business. Between 1940 and 1970, 4.5 million blacks migrated to northern and western cities. From 1910 to 1970 the proportion of blacks living in metropolitan areas increased from 27 to 74 percent. With rural-to-urban migration, black landownership declined by 330,000 acres per year from 1954 to 1969 (see Figure 5.6). During the 1970s blacks lost land at a weekly rate of 6,000 acres in seven southern states. At that rate, no black landowners will be left by 1990 (see Figure 5.7).[22]

The loss of over 10 million acres of land since 1910 has had devastating consequences for blacks. Land constitutes the largest equity base under black control. It is more than a means of earning a living or providing psychological well-being; landownership is the basis of political strength. Landowning blacks are more likely to register, to vote, to participate in civil rights actions, and to run for public office than nonlandowners. "In effect, landownership in the rural South confers on blacks a measure of independence, of security and dignity and perhaps even power."[23] Landowning blacks represented a major force for the rural civil rights movements of the 1950s and early 1960s. They provided a consistent and solid electoral base for independent black political activism through organizations such as the Mississippi Freedom Democratic Party.[24] A statistical analysis of voting patterns in the twenty-nine black majority counties in Mississippi showed that the key to black voter turnout was not the *amount* of their income but the *source*. Counties with the highest turnouts were generally not the wealthiest but had sizable numbers of black landowners and other blacks whose incomes were not derived from whites. In low-turnout counties, wealthier blacks lacked a land base and were thus more dependent upon whites for their income.[25] The loss of black landownership, then, has meant the loss of political power for blacks.

Black landowners face a particular problem: heirs property. One-third of land held by blacks in the rural South can neither be bought, sold, or traded,

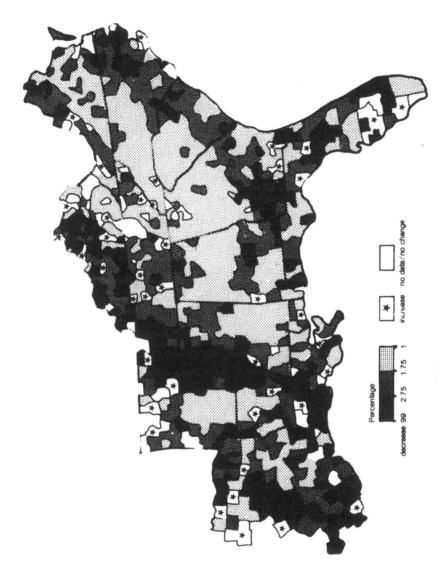

Figure 5.6. Black farmland change as a percentage of all farmland change, 1954–1974. *Source: Census of Agriculture,* various years.

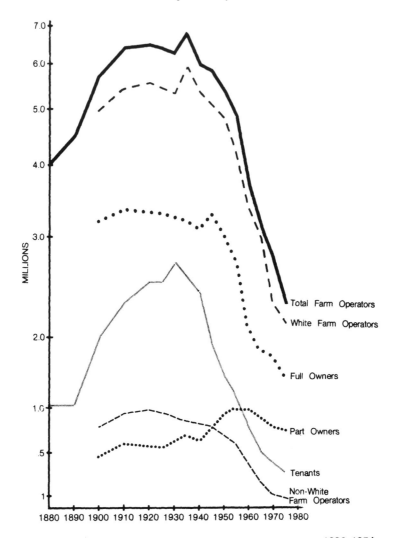

Figure 5.7. Changes in U.S. farm operators by tenure and race, 1880–1974.
Source: *Census of Agriculture.*

nor can it be used as collateral for agriculture or housing. Poor people in
general, and blacks in particular, have not been taught the importance of writing wills and seldom can afford an attorney to write one. At death, then, the
dispensation of their property is subject to state laws that often recognize
multiple heirs. Under this practice, a farm is inherited by all of the survivors
in the immediate family, usually the children. Later, all of *their* children
inherit their parents' share, and within two or three generations a 160-acre
farm might be owned by 100 people!

Such complicated ownership very often contributes to black ownership
decline through resulting involuntary sales. For example, if one heir wants a

cash settlement instead of land, the courts can, and often do, force a sale on the entire plot, even though another heir currently farms the land and does not wish to sell. Family disputes are thus easily exploited to require small farmers to sell out to larger expanding operations. Such involuntary sales remain a serious problem throughout the South, especially in Alabama and Louisiana.[26] In one rural Alabama county, for example, the probate judge owns 10,000 acres. The tax records show that he has title to only about half that acreage. He simply includes the rest in his personal holdings and enters it on his tax records. A good portion of this land legally belongs to blacks who have had to abandon it, usually because of economic or inheritance-related reasons, or to black heirs who have not been informed of their inheritance.[27]

Multiple ownership also frequently results in tax sales. Usually one heir, the one living on the land, pays the property taxes for all of the owners. But when the resident heir dies, there is often no nonresident owner who is willing or even able to assume the tax burden for all of the other heirs. As a consequence, they are forced to sell their land.

In addition to the likelihood that heirs property will have to be sold, multiple owners who are able to keep their property find that it is less valuable than single-owner property and can seldom be used as collateral. Even though the Housing Act of 1949 does not insist on clear title but only security upon the applicant's equity in the farm, the Farmers' Home Administration (FmHA) rarely lends money to owners of heirs property.

Although blacks suffer the same problems as small farmers everywhere in the country, disproportionate numbers of them are among the smallest and the poorest and are therefore particularly vulnerable to the competition of agribusiness. Three alternative explanations exist to account for this vulnerability. The conservative explanation, the most common, blames individuals for their failure to adapt to trends in agriculture in the United States. For example, according to James Fisher, a geographer: "He [the black farmer] commonly has a less advanced attitude towards improved methods and commercial production, and therefore does not use his land as effectively as he might. The inability of Negro farm owners to adapt is reflected also in limited commercialism and social attitudes and his own resistance have virtually excluded him from participation in the administration of the programs (soil conservation, farm loan) which vitally affect him."[28] These explanations lay the problem of black farmland loss at the feet of blacks themselves. The social and economic context in which these farmers exist is ignored. This conservative approach is particularly attractive because whites and white institutions are not held accountable for the problem and are allowed to feel sorry for this "unfortunate" minority, as individual traits of adaptation and intelligence can hardly be changed by society.

The liberal perspective explains the decline of black farmland as the result of racist attitudes, behavior, and institutions in the United States. The prejudices and practices of white racism succeeded in excluding blacks, for the most part, from participation in the free nineteenth-century land-granting programs of the federal government. Thus, while poor whites became landed farmers, blacks were left as sharecroppers on the plantations where they had formerly been slaves. Even those who did acquire land of their own were denied the social and economic resources necessary to escape poverty and illiteracy. Unfortunately, very little documentation is available of the racist barriers black farmers have faced and continue to face. The present legal and economic difficulties associated with the remaining black farmland owners stem directly from these historical and contemporary experiences.

The radical paradigm, the least known and developed, views discrimination and the demise of black farm families as inevitable in the present economic system. Capitalism permits greater profits to be extracted from black farm operators and pits white and black farmers against each other, thus weakening white farmers as well. Without adequately sized farms, both races are forced out of farming to provide cheap labor to urban employers and, especially in the South and Southwest, to provide inexpensive land and labor to remaining agriculturalists operating on a larger scale.[29] The two great labor migrations to northern industrial centers, blacks from throughout the South and whites from the Appalachian highlands, were direct results of farms that were too small to provide an adequate living.

In general, radicals do not believe that capitalism is the originator of racism, but rather that capitalism perpetuates and intensifies racism because of the valuable function it serves. Similarly, the demise of small family farms provides a labor force that can easily be exploited. Farm incomes, regardless of race, are substantially below those in commerce and industry.[30] Greater exploitation of farm labor means higher profits for agricultural input (fertilizers, machinery, bank loans) and output (grain merchants, livestock markets) industries and lower food costs to urban consumers. Thus commercial interests and consumer groups have a vested interest in maintaining the exploitation of farm labor—both black and white.

Radical analysts, who consider not only class and race antagonisms but also consequent regional and urban/rural inequalities, employ the concept of regional colonialism or regional underdevelopment.[31] Andre Gorz has suggested that

> the geographical concentration of the process of capitalist accumulation has necessarily gone hand-in-hand with the relative—or even absolute—impoverishment of other regions. These latter regions have been used by industrial and financial centers (higher-order urban clusters) as reservoirs of labor or of primary and agricultural resources. Like the colonies of the great European empires, the "peripheral" regions have provided the metropoles with their savings, their labor-power, their men without having a right to the local reinvestment of the capital accumulated through their activity.[32]

Black farm families are impacted as both a marginal race and a marginal occupation. Black farmers experience, along with most other farmers (especially small-scale operators) in the United States, the marginalization of their lives in an urban, industrial market exchange economy; but they experience it to a greater degree. Victims first of agricultural capital, later of industrial capital, and constantly of personal and institutional racism, rural blacks predictably have lost their land base.

Conclusion

The U.S. government has been quick to recognize the link between people's well-being and their control over land resources in Third World countries. There, say Western politicians and scientists, absentee landlords and rural poverty must obviously be replaced by land reform and redistribution of wealth. But in the United States the same issues are cloaked in less clarifying terms like "land use," "conservation of resources," and "protecting farm-land."[33] Indeed, advocates of land reform in the United States are generally considered radical, even though the concept of land reform is not alien. Most of the late-nineteenth- and early-twentieth-century land-granting policies were based on the ideals of land reform. U.S. foreign policy also embraced land-reform programs when they were imposed on Japan after World War II and on South Vietnam in the 1960s.

With 5.5 percent of all farms in 1974 controlling more than 50 percent of all farmland,[34] even conservative agricultural economists such as Don Paarl-berg, former economic adviser to President Nixon's Secretary of Agriculture Earl Butz, recognized the serious problem of current trends in U.S. farmland concentration: "This nation is in danger of developing a wealthy, hereditary land-owning class of people not unlike that of other nations. . . . Inheritance will be the only way for young people to get into farming."[35] One of the dangers of concentrating so much power in a few hereditary hands is that landownership carries with it social and political power. Yet influential policy-makers are oblivious to these antidemocratic trends, preferring instead to serve the interests of agribusiness. President Carter's Secretary of Agriculture, Bob Bergland, reflects this perspective: "If land values declined, or producers proved unable to service their debts for any reason, foreclosures could be expected. Under these circumstances, *concentration of land ownership would seem inevitable*" (italics added). In Catch-22 fashion he adds, "But if land values continue to escalate, owner equity will continue to rise, encouraging aggressive operators to continue expanding. With this alternative, *further concentration of land ownership also seems inevitable*" (italics added).[36] Clearly, despite their attempts to preserve the myth of the family farm, U.S. politicians accept the "inevitability" of its death.

Part 3

The Market Economy and Agribusiness

The Myth of Large-Scale Efficiency

The danger of measurement is that we will regard it as truth instead of as evidence.

—Kenneth Boulding

U.S. agriculture *seems* efficient. From 1950 to 1971 agricultural production increased 52 percent and output per farmer increased 257 percent. This was accomplished by fewer farmers working larger farms. Total farm labor decreased from 11 million workers in 1945 to about 4 million in 1977, and average farm size increased from 210 acres in 1950 to around 400 acres in 1977. The number of farms of 500 acres or more increased while farms of less than 180 acres fell sharply. Furthermore, Americans spend an average of 17 percent of their disposable income on food, one of the lowest percentages of any country (Figure 6.1). American agriculture must be efficient . . . or is it?

How are food production and farm size related to efficiency? To answer this question, terms like "efficiency," "production," and "economies of size" must be defined and the supporting evidence for efficiency in agriculture examined.

Agricultural economists define *technical efficiency* as the greatest output for any set of inputs in a given period of time. Because farming consists of several components—producing food and fiber, purchasing agricultural supplies, and marketing produce—farmers can be efficient in any one or all of these activities. The term *economic efficiency* applies to the whole farm operation rather than to only one activity. Farmers might be technically efficient in producing food, but inefficient in, say, marketing, and thus overall they might be economically inefficient. In contrast, farmers might be extremely efficient at purchasing inputs and marketing outputs but not at growing crops; yet their overall performance might be considered efficient. The former case describes many family farmers, the latter case that of large-scale producers, especially industrial-type farmers.

Increased *production* is frequently equated with efficiency. Data show that productivity of U.S. agriculture increased rapidly after the 1930s (but less so since 1975). From the 1950s to the 1960s crop yields, for example,

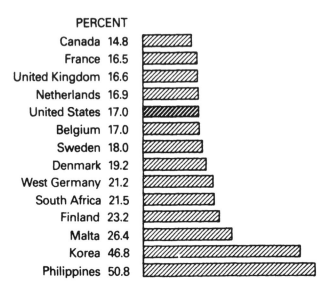

PERCENT

Canada	14.8
France	16.5
United Kingdom	16.6
Netherlands	16.9
United States	17.0
Belgium	17.0
Sweden	18.0
Denmark	19.2
West Germany	21.2
South Africa	21.5
Finland	23.2
Malta	26.4
Korea	46.8
Philippines	50.8

Figure 6.1. Share of after-tax income spent on food in various
countries, 1975.

increased 60 percent for cotton, 58 percent for corn, 47 percent for wheat,
and 22 percent for hay. In the aggregate, productivity in farming, measured
by an index set at 100 in 1967, went from 53 in 1910 to 113 in 1975.[1]
USDA officials are proud to point out that modern U.S. agricultural systems
allow one farmer to take care of 75,000 chickens, feed 5,000 head of cattle,
and milk 60 cows. For proponents of agribusiness the ultimate statistic for
efficiency is that one U.S. farmer feeds forty-eight persons.[2]

Efficiency and greater production are not the same. Large-scale farmers,
measured in number of acres, produce more agricultural goods and higher
yields per worker. Yet all published data indicate that the larger the farm
operation, the lower the *yields per acre*. Proponents of agribusiness measure
the success of farming by using output per farmer, but this index confuses
agricultural efficiency with productivity. Large-scale farmers *seem* more efficient, but they actually only produce more in quantity and value with fewer
workers and more land. According to Luther Tweeten, an academic agricultural economist, the cost of all inputs (including the opportunity cost of
equity capital and of operator and family labor) per unit of output (including
receipts from farm commodities, nonmoney income, and government payments) favored the largest farm size categories, measured in gross value of
farm sales. Farmers with sales under $25,000, on the average, lost money and
did not cover all of their production costs in 1960. Farmers with sales of
$25,000 and over cleared a profit and that profit increased with gross sales.[3]
But in terms of efficiency, Tweeten concluded, "Most of the economies of
size appear to be achieved by Class II, farms and unit costs decline very
slowly beyond an annual output of $30,000 per farm"[4] (see Figure 6.2).

Tweeten noted that small-scale farmers earned low returns because they

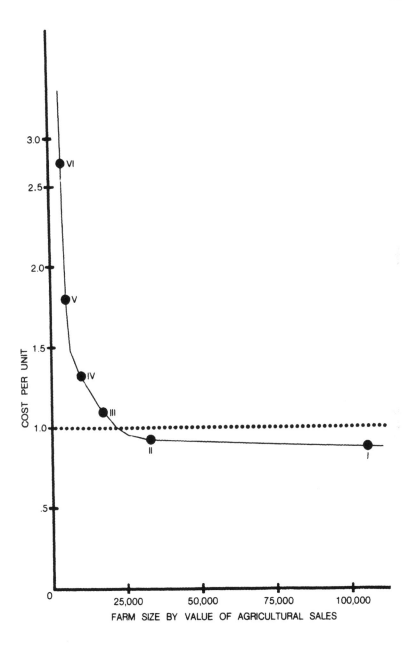

Figure 6.2. Long-run unit cost of farm production, by economic class of farms, 1960. The marginal and average revenue curves are indicated by the horizontal dotted line at $1. *Source:* Luther G. Tweeten, "Theories Explaining the Persistence of Low Resource Returns in a Growing Farm Economy," *American Journal of Agricultural Economics* 51, no. 4 (1969), p. 810.

had to pay too much for their land. "There is constant pressure to expand farm acreage to achieve economies of size."[5] Savings from greater efficiency are bid into the price of farmland. Potential buyers who are unwilling or unable to pay this price find that land is bid away from them by investors who can afford the initial investments that will make profit maximization possible.

1971 data confirm that farms with gross sales of $20,000 to $40,000 are the most efficient while farms with over $40,000 sales are the least efficient, even though they have large gross sales.[6] The largest farms also have the highest proportional production expenses relative to the acres cultivated and net income realized (Figure 6.3). Similarly, the *1974 Census of Agriculture* showed that small-scale farms with gross sales between $20,000 and $40,000 produced a margin of profit comparable to medium-scale farms with gross sales between $40,000 and $100,000 and a better margin of profit than large-scale farms with $100,000 and more of farm sales. In other words, the smaller, full-time farmers *must* be efficient or they would not survive, while larger farmers can afford to be and indeed often are inefficient.

By considering only yields per farm worker, all other workers who contribute to food production (and without whom farmers could not be as productive) are excluded. One out of every seven U.S. civilian workers manufactures farm inputs—such as machinery, fertilizers, pesticides, and fuels—

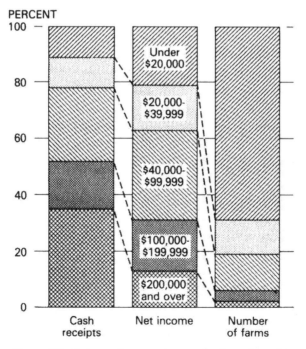

Figure 6.3. Cash receipts, net income, and number of farms by sales classes, 1977.

or processes and markets agricultural products, such as food, leather, and industrial oils. It takes far more than one farmer to feed forty-eight people; the farmer's contribution alone actually feeds only sixteen.[7] Just as efficiency statistics are inflated by an underestimation of labor input, they also distort the amount of capital input by American consumers. The claim that Americans have to contribute only 17 percent of their disposable income to agriculture must be seen as an average in the context of a society with an incredible and increasing income gap between rich and poor. While families with an income of $25,000 (not even particularly "rich") spent only 14 percent on food in 1973, families with incomes of less than $3,500 spent 79 percent.[8] Thus, the most accurate measure of farm productivity is one that compares output with the combined use of *all* resources—not just farm labor, but also industrial labor, capital, and energy.[9]

Economies of Size

Farm size is a critical variable in measuring efficiency. What size, large or small, provides the best and least costly delivery of agricultural products? This question must be answered as it relates to individual farmers, to the agricultural sector in general, and indeed to the society as a whole. In the popular and scientific mind, efficiency is usually associated with large-scale farm operations and inefficiency with small-scale farms. In the late 1930s, the number of all farms peaked at 6.8 million; by 1977 farms numbered 2.7 million. Since small-scale farms declined sharply and large-scale farms increased, apologists for agribusiness agree that inefficient producers have been weeded out and the efficient ones have been rewarded. But what are the *facts* on the economies of farm size and efficiency?

The term "economies of size" refers to the reduced production costs and increased returns that are achieved when farmers expand their operations.[10] Fixed costs can be spread over more and more units of output, resulting in lower costs per unit. But the economies of size do not increase with unlimited expansion. J. Patrick Madden, an agricultural economist, has made the only major review of the literature on farm size and efficiency.[11] Examining studies on seven types of crop farming under fourteen different circumstances in five different states, he concluded that "in most of these studies, all the economies of size could be obtained by modern and fully mechanized 1-man or 2-man farms."[12] The acreage involved in a one- or two-person farm varied greatly, as Table 6.1 shows. For example, fresh vegetable farms of 640 acres were almost as efficient as farms of 2,400 acres and over, if adequate field workers were available. Some large-scale farms are listed as most efficient in the Imperial Valley, for example, simply because in those areas there were no small-scale operations for comparison. Where the comparison was possible, fully equipped, small-scale farmers achieved nearly all of the technical economies of size.

Increasing farm size beyond the two-person family farm is not justified on the basis of efficient resource use. Hence, to say that costs per unit of output decrease as farm size increases is true only to a very limited degree. Yet

TABLE 6.1
A Sample of Economies of Farm Size in the 1950s and 1960s

Type of Farm and Location	Farm Size with Lowest Average Costs
Crop Farms	
California	One-Person Farm
cling peaches	60-tree orchard, if not mechanized
	90-110-tree orchard, if mechanized
Texas High Plains	
irrigated cotton	440 acres, with 102 acres of cotton, 6-row machinery
Columbia Basin, Oregon	
wheat	1,000 acres
Southern Iowa	Two-Person Farm
cash grain	320-360 acres with 3-plow tractor
Fresno County, California	Four-Person Farm
irrigated cotton	700 acres in heavy soils and 1,400 acres in light soils
California	Unspecified-Person Farm
Imperial Valley	
field crop	1,500-2,000 acres
vegetables	640 acres with timely contract services
Kern County	
cash crops	640 acres (160- and 320-acres almost as efficient)
Yolo County	
cash crops	600-800 acres
Feedlots	
Colorado	1,500-head capacity
Imperial Valley, California	5,000-head capacity
Dairy Farms	
New England	One-Person Farm
	35 cows
Minnesota	48 cows, 290 acres, 3-plow tractors, $160,000 investment
Iowa	Two-Person Farm
	32 cows
Arizona	Three-Person Farm
	150 cows

Source: J. Patrick Madden, Economies of Size in Farming, USDA, Economic Research Service, Agricultural Economic Report No. 107 (Washington, D.C.: U.S. Government Printing Office, 1967).

proponents of large-scale agriculture counter by referring to the "survivor technique": farm sizes that are efficient will survive and farm sizes that are inefficient will decline.[13] Since the number of small-scale farms is decreasing and the number of large-scale farms is increasing, large-scale farms must be the efficient producers. The survivor argument is deceptive because farmers may survive for many reasons other than their efficiency. Large-scale producers receive disproportionate advantages from the government in payments, tax and labor laws, buying farm inputs, and marketing products. According to land economist Philip Raup, "Economists perpetuate the myth that one dollar is like another dollar, in calculating costs or prospective rates of return."[14] Payments by the federal government, for example, reduce uncertainty in farming, "quite literally a form of guaranteed annual income." The greatest beneficiaries are large-scale farmers with substantial indebtedness. Large-scale farmers then become more competitive not because they are efficient but because they have greater profitability.

Increasing Size, Increasing Profits

For the Texas High Plains, the most efficient irrigated cotton farm is a one-person operation of 400 acres of cropland.[15] However, census data show that precisely this size category, less than 500 acres, decreased from 1959 to 1964. The next category, 500 to 999 acres, remained essentially constant. Farms from 1,000 to 1,999 acres—the least efficient—showed an almost 5 percent increase in their share of the harvested cropland.

Why do farmers expand beyond the maximum unit of efficiency? In the United States, conventional wisdom states that increased profitability is synonymous with greater efficiency. In reality, profits increase with farm size, irrespective of efficiency, which eventually declines. In the Texas example, farms with 1,000 to 2,000 acres of harvested cropland required a three- to five-person regular labor force, plus one million dollars of investment. "Compared with a one-man 440-acre operation, these larger farms seem to have slightly higher average total cost, but because of volume, their total profits are substantially larger."[16] Figure 6.4 illustrates that as efficiency levels off, or even decreases slightly, total net profits continue to increase. But small-scale farmers reach production efficiency before they can maximize profits. J. P. Madden found this irony to be true for cash-grain and crop-livestock farms in Iowa, irrigated cotton farms in California, feedlots in California and Colorado, and dairy farms in Iowa and Minnesota: "The Minnesota study indicated that the 1-man dairy farm could realize little, if any, increase in efficiency by doubling farm size and hiring an additional worker, but the increase in volume would give rise to considerably higher profits."[17]

D. K. Britton and Berkeley Hill, British land economists, provided a simple arithmetical example showing that the two objectives of efficiency and profitability are contradictory.[18] In terms of efficiency, it is better to achieve an output of $12,000 from an input of $10,000 than it is to achieve an output of $23,000 from an input of $20,000. The first example is more efficient since every dollar of input yields $1.20 of output. In the second case, each dollar yields only $1.15. But most farmers would certainly prefer the second situation since it leaves them with a balance of $3,000—a 50 percent improvement on the balance of $2,000 resulting from the first situation. This example illustrates that large-scale operations can be less efficient with every unit of additional input than medium- and small-scale farms would be if they employed these additional units. A USDA study using data for 1970 found that the rate of return to equity capital ranged from –6.1 percent for farms with sales of less than $2,500 to 6.9 percent for farms with sales of $100,000 and

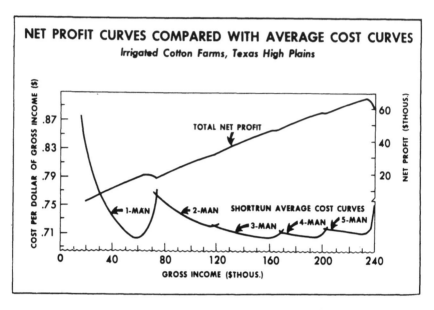

Figure 6.4. *Source:* J. Patrick Madden, *Economies of Size in Farming*, USDA, Economic Research Service, Agricultural Economic Report No. 107 (Washington, D.C.: Government Printing Office, 1967), figure 2, p. 20.

more. If returns from land appreciation are included, then the rates increased to −3.2 percent and 10.8 percent, respectively.[19] Thus, as long as large-scale producers can increase their profits, it is in their self-interest to expand, even if this means becoming relatively inefficient. Hence, what is rational and beneficial for individual farm operators (profits) is irrational and wasteful for the society as a whole (inefficiency in resource allocations).

Family farmers, in particular, face a further problem in the efficiency-profitability squeeze. In his presidential address to the American Agricultural Economics Association, B. F. Stanton outlined the problem: "A farm family tries to get the most it can out of the bundle of resources it controls. And it is not net income by itself that matters to a family. Rather, it is some larger combination of things including survival, net income over time, enlarging the bundle of resources [for the next generation] that the family controls, and increased prestige within the local social system."[20] Family farmers, especially in contrast to large-scale industrial farm operators, are motivated by the desire to achieve the ideals of family farming. These very positive human goals are exploited under present conditions of agrarian capitalism to the detriment of family farmers themselves, who achieve efficiency without the profit they deserve.

Ultimately, family farming can be viable only as long as farm income—that is, cash receipts for products—exceeds production expenses by an amount sufficient to produce a livelihood to those who work the farm. For the first half of the twentieth century, the net income of farmers amounted to about 40 percent of gross receipts. Since 1950 the profitability of farming has declined sharply. Net income was only 29 percent of gross income in 1972. During this recent period, production expenses rose more rapidly than prices received by farmers for their products did. This cost-price squeeze meant that farmers with limited resources, mainly tenants and family farmers, had a harder time providing themselves with an adequate livelihood. As long as farmers are committed to farming as a way of life, then, they will have to increase their production to remain viable in farming as a way of business.[21]

Farmers are successfully increasing production and profit, though not efficiency, by increasing size. In California the net income for a 260-acre farm went from $5,877 in 1950 to $7,355 in 1969, an actual decrease in constant dollars considering inflation. On the other hand, farmers who increased to 617 acres realized a net income of $17,455 in 1969.[22] The conclusion is inescapable: farmers must increase their production by acquiring more land, buying more and larger machinery, using more fertilizers, and hiring more farm workers. Farmers are involuntarily caught on a cost-price treadmill. Oligopolistic agribusiness input and processing firms squeeze farmers from both directions, and the only direction farmers have to go is up in size: get bigger or get out. The myth of large-scale efficiency makes it appear that individual farmers are responding to the "free market." In reality they are involuntary victims of agrarian capitalism.

Intensification of farming and economies of scale are also frequently confused, as both result in increased production. Farming is intensified if the same acreage is used to produce higher yields per acre. Most commonly, this is achieved through greater inputs of capital in the form of fertilizers, pesticides, and hybrid plants and animals. But intensification can also be achieved through better use of labor; for example, dairy farmers can increase their milk production by 10 to 20 percent if they switch milking from two to three times a day.[23] Family farmers could intensify their production if they had technology appropriate for their labor needs and management skills to maximize their resources.

As Figure 6.5 shows, increasing the size of any farm, regardless of its level of intensification, will achieve economies of size and increase production (A to A' and B to B'). The benefits, however, do not increase indefinitely. Often, beyond a certain point (A' and B'), which is usually much smaller than the current farm sizes suggest, production levels eventually decrease as diseconomies of size set in. The only farms that are really forced to increase their size are those that are as fully intensified as possible. When farms are fully capitalized, the only way to achieve economies of size is to increase acreage. But in reality, family farmers are undercapitalized and could significantly increase their production by intensifying. Thus, a farm at point A could increase production equally by increasing size (C) *or* by intensifying (B). At any size category farmers may be undercapitalized, so that large-scale farmers could also improve their performances by moving from A' to B', which represents the maximum level of useful capitalization under current levels of technology.

USDA officials and agribusiness advocates seem to Americans to believe that agricultural "progress" can be achieved only when farmers increase pro-

duction by increasing size (i.e., moving from A to A'). This assumes tha
farms at A and A' have reached maximum intensification. However, increase
production can also be achieved with less displacement of small-scale farmer
by assuring that sufficient factors of intensification are available to farmer
at A. By making farming at A more intensive, farmers could increase outpu
to B. Indeed, smaller, more intensive farms would produce the same outpu
as much larger but less intensive farms. Since more farmers are toward the A
end of the curve than near A', the intensification approach would improv
the relative standing of most farmers. In contrast, the "get bigger" approacl
enhances the position only of those who can afford to get bigger, to the detri
ment of those who cannot.

Agribusiness proponents often argue that small-scale farms are "inefficient'
because they cannot accommodate the huge acreage demands of the new pro
duction technology. But as Virgil L. Christian, Jr., and Carl C. Erwin hav
pointed out, "It is one thing to say that small farmers were rapidly displace
because agricultural production became more capital intensive. It is quit
another thing to say that the changes in production methods created economie
of size that made the small farmer's position untenable even if his holding wa
large enough to accommodate the smallest unit of technologically efficien
capital."[24] The thrust of publicly funded agricultural research in the Unite
States is toward technological improvements that demand large-scale farm

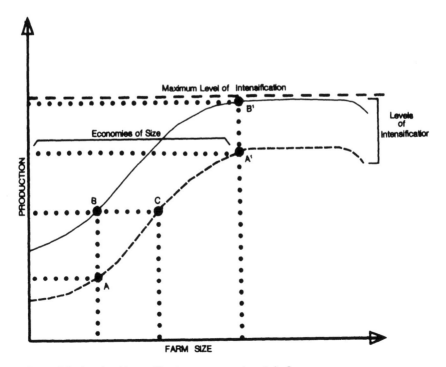

Figure 6.5. Levels of intensification or economies of size?

to reach economies of size and render small-scale farms unviable. The government could just as easily spend our resources on development of intensification technology that would strengthen family farms.[25]

There are, then, two approaches to achieving the goal of maximum efficiency under present technological conditions. Proponents of agribusiness argue that if farms get larger they will automatically become more efficient; proponents of family farms argue that by increasing farm intensification farms will become more efficient. The former approach has the disadvantage of concentrating landownership, wealth, and political power in fewer hands, displacing more farm families, and not increasing efficiency (indeed, in some cases, actually lowering efficiency). The latter approach has the advantage of displacing fewer farm families, sharing rural decision making and wealth widely, and increasing total efficiency and output.

Economies of Size in Purchasing and Marketing

Largeness in and of itself is advantageous in the marketplace. Large-scale producers receive discounts for bulk purchases of inputs, such as fertilizers, insecticides, feed, and fuel (see Table 6.2). Even the U.S. Chamber of Commerce, one of the strongest champions of agribusiness, admits, "Usually these successful firms [very large farms] have been able to buy many of their inputs, except labor and management, for from 15 to 25 percent less than the price paid by more small units."[26] Discounts are justified because large orders reduce administrative and transport costs per unit. Furthermore, in spite of charging a smaller percentage profit margin per unit, high-volume sales yield a larger total profit for agribusiness input firms.

Under current conditions, farmers must continually borrow money to

TABLE 6.2
Relationship Between Farm Size and Cost of Capital and Other Purchased Inputs

Farm Size (acres)	Interest on Operating Capital[a] (Percent)	Percentage Volume Discounts		
		Fertilizers	Insecticides	Crop Dusting & Aerial Spraying
80	6.88	0	0	0
160	6.52	4	0	0
320	6.47	4	5	0
640	6.47	4	5	12.5
1,280	6.15	10	8.5	17.5
3,200	5.90	10	14[b]	25

[a]Six percent was the norm.
[b]Only one observation.

Source: J. E. Faris and D. L. Armstrong, Economies Associated with Farm Size, Kern County Cash Crop Farms (Washington, D.C.: Giannini Foundation Research Report No. 269, 1963), pp. 73-96.

expand production. Again, large-scale farmers have an advantage. Table 6.2 shows that as farm size increases, interest rates charged for credit drop. Because large-scale farmers have larger assets, they have greater credit worthiness for banks than the lower-valued small-scale farmers. Smaller farmers have the added handicap of being less likely to negotiate a bank loan in the first place and are forced to enter more purchase agreements for machinery and other farm supplies with agribusiness firms, which offer credit when banks will not, but often at higher interest rates. The viewpoint of input agribusiness is presented by the president of John Deere and Company: "To use credit is a sales tool. We provide it because we must [because banks do not]. . . . The paper we accept from our dealers carry higher rates than the banks charge for such paper. . . . [T]he limited availability of credit from other lower cost sources must be a factor in the situation."[27] Thus, the more family farmers try to catch up with large-scale farmers by expanding, the more they get behind in debt.

Large-scale farmers have further advantages in marketing their produce. For example, big farmers may receive high prices for grain because they have the facilities to store grain on their farms beyond the period immediately after harvest, when prices are usually low, until a time when prices are higher. Large-scale farmers bypass not only local stores when they purchase inputs, but also local market outlets when they market their crops and livestock. Large corn producers in the middle West, for example, market "directly to a [grain] terminal or to large-volume users."[28] Iowa farmers producing over 300,000 bushels of corn increased their effective selling price by 5 cents per bushel and more by using their own elevator facilities and by selling in large volumes directly to riverside terminals. "Together with the $10 per acre advantage in purchased inputs, this adds up to a sizable net income advantage of about $15 per acre of large-scale operations."[29]

Social Diseconomies as a Result of Profit Maximization

Studies indicate that long-run average cost curves for farming are not U-shaped (suggesting economies of size up to a certain farm size followed by diseconomies), but rather L-shaped (implying economies of size up to a certain point followed by neither diseconomies nor economies as production units become larger). However, although the L-shaped curve holds true for individual farm operators, at least up to $100,000 gross sales, the U-shaped curve is most appropriate for societal outcomes of increasing farm size (see Figure 6.6). That is, very large farms may not have diseconomies because they can externalize many of their costs to society as a whole. The costs of urban unemployment, caused by displacement of farm families and grossly underpaid migrant workers, and of water and soil pollution—to name only a few related problems—must be borne by taxpayers at large.

As Figure 6.6 shows, the optimal size of farms is much greater for individuals than it is for society. If efficiency is the goal, a farm can expand until it reaches economies of size (A'). If profitability is the goal, it can continue to expand until a theoretical point where diseconomies may set in (B'). But such

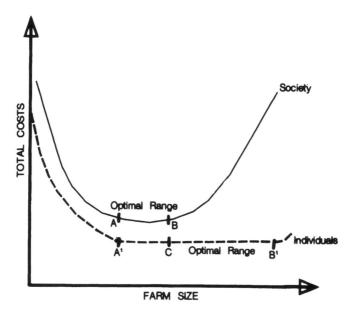

Figure 6.6. Optimal range of farm sizes for individuals and for society.

great expansion is only possible because at a certain point (C) large-scale farmers begin to externalize their diseconomies while simultaneously receiving government subsidies, credit discounts, and tax advantages—all of which artificially encourage growth toward B'. In terms of societal costs, the optimal farm size is only from the point of economies of size (A) to the point of social diseconomies of size (B) and not beyond. Enlarging units of production for profitability alone concentrates wealth and political power in fewer hands and results in more people being dependent on a small elite for their livelihood and freedom. The profit advantages of large-scale producers, then, result directly in societal diseconomies. *Large-scale farms should more properly be called over-sized.*

In summary, large-scale and small-scale farmers are efficient in different ways. Each way benefits and harms different groups. On one hand, large-scale producers are efficient because capital-intensive mechanization replaces labor, which makes higher yields per worker possible. These farmers are also efficient at making profits, which is accomplished by disregarding social and environmental costs of their production. Once economies of size are reached, however, large-scale farmers are no more efficient in terms of input costs per unit of output than family farmers. On the other hand, small-scale farmers are efficient because they have higher yields per acre with their labor-intensive production methods. By not displacing as much labor with machines, family farmers provide greater employment opportunities under capitalism than large-scale farmers do. Because small-scale producers reap lower profits, they also create fewer social and environmental problems for society. Collectively, they allocate societal resources more rationally than large-scale producers.

A rational agricultural policy would facilitate farmers' expansion up to point *A* and inhibit their growth beyond *B* (as in Figure 6.6). Most economists have not proposed rational farm policies because, as scientists in the logical positivism tradition, they concern themselves only with variables that are directly measurable and accountable (in a cause-and-effect relationship) and they conveniently forget or ignore variables not chosen for study. As society pays the price for the externalized costs of its institutions, isolating the portion related to the social diseconomies of large-scale farming is not easy empirically. Thus, economists study the L-shaped curve and ignore the U-shaped reality. Furthermore, ideologically most scientists have sided with agribusiness in the use of neoclassical economic models. The selection of their theoretical framework—agrarian capitalism rather than agrarian democracy—allows them to ignore relevant human factors in the expansion of farm sizes. The concept of efficiency has thus been measured and interpreted in its narrowest form: to achieve short-run, individual gains. The concept itself is valuable, but its application has been distorted and used against, rather than for, family farmers.

Social Justice and Economies of Size

In conclusion, the myth of large-scale efficiency provides a justification for the massive displacement of rural people and for getting the remaining farm families to produce even more food for less income. From 1950 to 1976 the parity ratio (prices received divided by prices paid by farmers) declined by 40 percent.[30] The relative increase in net farm incomes for large-scale producers in particular has obscured the absolute increase in exploitation of farmers in general. Efficient producers often receive inadequate income from their work. Instead of rewarding efficiency in production, market forces and federal farm programs and tax laws reward inefficient producers because they are sufficiently large to be profitable. In the United States, *profitability is the actual measure of success.* Using "efficiency" to stand for "profitability" hides the negative consequences of agribusiness and reduces public criticism of agrarian capitalism.

Technical efficiency, which is a measure of capital, does not address the question of social justice, a measure of labor. For individuals and society as a whole, a slightly "inefficient" agricultural sector, in the technical sense, may achieve higher social justice and thus increase the overall "efficiency" of the national economy—fewer people on welfare in cities, less unemployment, less malnutrition, and less concentration of economic and political power. *Both efficiency and social justice can be achieved in U.S. agriculture.* By truly rewarding efficiency, family farmers will be strengthened; by discouraging inefficiency, large-scale producers will be reduced. The two goals of efficient allocation of national agricultural resources and survival and prosperity of family farmers are the same. Large-scale farmers, on the other hand, stand in conflict with national economic and democratic goals.[31] From the standpoint of these goals, discussion of large-scale efficiency may be a matter of "finding out the best way of doing something that should not be done at

Figure 6.7. A farm family displaced by agribusiness in search of a new home. Tulelake, California, 1939. Dorothea Lange, FSA photo.

all."[32] Poet Wendell Berry stated the choice this way: "The standard of the exploiter is efficiency, the standard of the nurturer is care. The exploiter's goal is money, profit; the nurturer's goal is health—his land's health, his own, his family's, his community's, and his country's."[33]

The Business of Agribusiness

The agricultural world and the industrial world are not two separate economies having merely a buyer-seller relationship. Rather, they are so intertwined and inseparably bound together that one must think of them jointly if there is to be any sound thinking about either one or the other.

—T. V. Houser
chairman of the board, Sears, Roebuck and Co.

Traditionally, family farmers once supplied many of their own inputs: they repaired machinery, mixed feed, and used their own manure. They also stored, processed, and distributed agricultural products. Both these input and output functions have been eclipsed by specialized agribusiness firms, leaving only agricultural production to farmers.

Agribusiness is the interrelated and coordinated food and fiber system. In its conventional definition, the concept includes (1) farm input industries providing services and machinery, fertilizers, pesticides, feed, and other supplies, (2) all commercial farmers, and (3) food-processing and -marketing firms.[1] J. H. Davis and R. A. Goldberg of the Harvard Business School first coined the concept in 1957. In 1967 they estimated that the value of sales from farm-supply firms to farmers totaled about $31 billion or about 33 percent of the total value of domestically produced food purchased by U.S. consumers. The farm sector itself added $12 billion or about 12 percent of the final retail value. The farm-product marketing sector added the remaining $52 billion or 55 percent of the total retail food value.[2] These data show that agricultural input and food-processing and -marketing firms dominated agribusiness already in the 1960s. In 1978, farmers received only 32 cents of every dollar spent by food consumers, and their share is continuing to decline (see Figure 7.1). From the 1930s to 1970s, the value of inputs into farming from the nonfarm sector increased faster than value added by the farm production sector, but the farm-product processing and marketing sectors grew the fastest of all three segments of agribusiness.

In this book, agribusiness is used in a slightly more restrictive sense than Davis and Goldberg used the term. All farmers produce some goods and

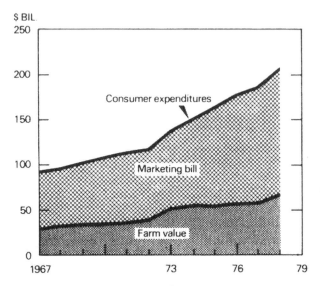

Figure 7.1. Farm value of total consumer expenditures for food, 1967–1978.

services for the market, but large-scale producers are most intimately tied to other branches of agribusiness; medium- and small-scale producers are more victims than beneficiaries of agribusiness institutions.

Agribusiness is dominated by large input and processing firms. In 1979, 15 companies accounted for 60 percent of all farm inputs, 49 firms did 68 percent of all food processing, and 44 companies received 77 percent of all wholesale and retail food revenues. (Figures 7.2 and 7.3 show the declining farm value of bread and tobacco.) The biggest agribusiness stockowner in the U.S. was Morgan Guaranty Trust, a top shareholder in 32 of the 157 agribusiness companies for which stock ownership was listed. Conglomerates, companies that own agricultural operations as well as unrelated businesses, constitute a major part of agribusiness. For example, Boeing owns Granny Goose Potato Chips, Dow Chemical owns Bud Ante Lettuce, and Greyhound owns Armour Meat Company.[3] Since food input and purchasing firms are dominated by large companies, they reduce their administrative costs by dealing with large-scale producers. The concentration of power in agribusiness firms encourages the same trend in the farm production sector as well. To ensure maximum total profits, agribusiness has an incentive to create strong control over the farm production process to acquire an adequate and timely supply of uniform quality farm products for the food-processing industries.

Agribusiness has no president, no board of directors, and no central office. Agribusiness firms do not meet in smoke-filled rooms plotting the destruction of family farmers; but they do share a similar perspective—that short-term profit maximization is the key to success. In addition, interlocking director-

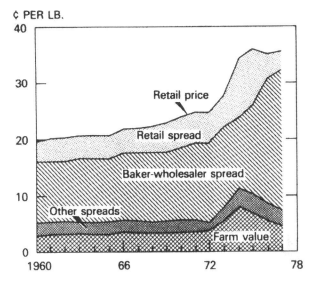

¢ PER LB.

Figure 7.2. Farm value of the retail price of bread, 1960–1977.

ships, where a person from one agribusiness board of directors sits on the board of another, intensifies this shared perspective and concentrates economic power in even fewer hands. The result is that *the business of agribusiness is profit*. For this profit to be socially acceptable (as it is usually acquired under noncompetitive market conditions and exploitative labor conditions) agribusiness relies heavily on the competing myths of the family farm and large-scale efficiency, using whichever one is most expedient to justify and disguise its exploitation of farmers and consumers.

Farm Input Oligopolies

All U.S. farmers are strongly tied to agribusiness. In 1973 farmers bought nearly $65 million of production inputs (see Table 7.1), three-fourths of which were purchased from farm-supply firms. The major items were feed, livestock, fertilizers, seed, and pesticides, as well as repair to buildings and machinery.[4] Farmers also purchased agricultural services, such as artificial insemination, livestock vaccinating, veterinary services, and farm management consulting.

Services

Farmers depend largely on a few agricultural service firms. According to the *1974 Census of Agriculture*, the largest farm service firms with $500,000 or more of gross sales took in 28 percent of all receipts but represented only 1.3 percent of all such establishments. The next highest category, from $100,000 to less than $500,000 gross sales, received 39 percent of all receipts and represented 12 percent of all firms. In total, 13 percent of the largest

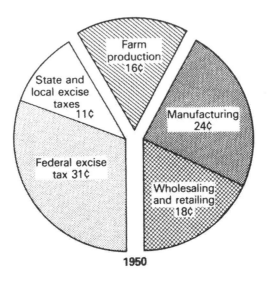

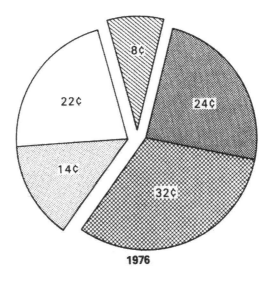

Figure 7.3. Declining farm share of each dollar spent on tobacco, 1950 and 1976.

TABLE 7.1
Major Farm Production Expenses for U.S. Farmers, 1973

		Millions of Dollars	Percentage
Purchased Feed		13,078	20
Purchased Livestock		8,152	13
Fertilizer and Lime		3,049	5
Seed		1,638	3
Pesticides		1,442	2
	Sub-Total	27,359	43
Repairs for Motor Vehicles and Farm Machinery		4,292	6
Repairs on Buildings		957	2
	Sub-Total	5,249	8
Labor		5,166	8
Interest on Borrowed Money		5,130	8
Rent to Non-Operator Landlords		4,152	6
Taxes on Property		3,321	5
	Sub-Total	12,603	19
Crop and Hail Insurance		465	1
All Other Expenses		13,904	21
Total Production Expenses		$64,746	100

Source: USDA, Economic Research Service, Farm Income Situation, FCS-44, February, 1973.

agricultural service firms grossed 68 percent of all receipts. Most of the critical supplies farmers need are controlled even more than the census indicates. A few giant agribusiness firms predominate. According to one study, in 1970 the four leading farm-supply firms controlled 67 percent of petroleum products, 71 percent of tires, 74 percent of chemicals, and 80 percent of rail transport.[5] When farmers become dependent on a small number of input firms it is the sellers, not the farmers, who create demand. This situation is called oligopoly.

The same oligopolistic conditions exist in the hybrid seed and tractor industries. Over one-half of the 68 million acres of corn in 1977 were planted with varieties from only two seed companies, Pioneer International HiBred and DeKalb. Pioneer controls slightly more than DeKalb, with 28 percent of the domestic corn market. DeKalb claims that over 20 million acres were planted with their hybrid corn products and sold through 7,000 farmer-dealers in the countryside. These two firms dominate the seed industry, which consists of twenty major seed corn companies, all corporate giants in their own right: Cargill, Trojan (a subsidiary of Pfizer Drugs), Ferry-Morse (a subsidiary of Purex), and Northrup-King (a subsidiary of Sandoz of Switzerland). These agribusinesses are very profitable, especially in contrast with the thousands of family farmers. Pioneer, the seed corn leader, showed 1975 sales from its corn business alone of over $124 million, over twice its sales just five years before. DeKalb reported a 22 percent increase in sales for 1977 over the first nine months' earnings of 1976.[6]

The nitrogen fertilizer industry, using natural gas as its most important raw material, also tends to be concentrated in the hands of a few large producers, and most of the large producers are either oil companies or have energy interests. In 1977, the eight largest producers controlled 41 percent of the production capacity for anhydrous ammonia. This oligopolistic control has allowed agribusiness firms to increase the price of this product beyond the increased costs of energy, and the profits have been extracted from farmers. Between 1970 and 1974, the cost of energy used in ammonia production increased by 49 percent while the cost of fertilizer went up 141 percent. Ammonia producers were receiving investment returns of 30 to 65 percent during 1973–1974 while the average U.S. industry received 11 percent and farmers received 4 percent.[7]

Farm machinery is also produced under oligopolistic conditions. In 1973, the four largest tractor manufacturers controlled 83 percent of the U.S. market. John Deere and Company and International Harvester, the top two manufacturers, controlled 57 percent of the total market. The sale of tractors with 80 horsepower and more, which includes the highly profitable supertractors, is even more concentrated. Together the two largest tractor companies claimed about 70 percent of the big-tractor market.[8] These agribusiness firms support the trend toward industrialized agriculture because this centralization means even more markets for their giant farm machinery. When tractors averaged 82.5 horsepower in 1973, Deere was producing 200 horsepower machines and talking of 500 horsepower giants[9] (see Figure 7.4).

Profits are the measure of the success of agribusiness. The average profit increase for all large corporations was 16 percent in the first quarter of 1974, but farm supply corporations did much better: International Harvester, 113 percent; Burlington Northern, 102 percent; Tenneco, 55 percent; Allis-Chalmers, 41 percent; and Ralston Purina, 32 percent. The average profit increase for the first quarter of 1974 for fifteen input firms was 66 percent.[10] Agribusiness used its oligopolistic power to extract wealth from farmers above and beyond the "normally accepted" competitive market price. According to U.S. Trade Commission statistics, the tractor industry overcharged U.S. farmers in 1972 by about $251 million. In the animal feed industry, dominated by Ralston Purina and Cargill, the overcharge in the same year was approximately $200 million.[11]

Farm input firms encourage large-scale farming and heavy reliance on purchased inputs through tax deductible research and "educational" expenditures. In 1970, agribusiness firms spent $174.1 million for research and development; and federal, state, and local governments spent $405.6 million. In addition, agribusiness companies spent about $150 million for technical educational assistance to farmers, an amount roughly equivalent to the educational expenditures of the Cooperative Extension Service at all of the land-grant colleges.[12]

Figure 7.4. Two kinds of farm technology. (Top) This futuristic combine will be able to harvest some 160 acres per eleven-hour day with 80 percent field efficiency or 310 acres per twenty-two-hour day with night operations completed at 75 percent field efficiency. Production will equal 40,000 bushels per eleven-hour day, or 77,500 bushels per twenty-two-hour day. Reprinted by permission of International Harvester. (Bottom) Family farm technology. Cotton picker attached to an all-purpose tractor, McLean, Texas, 1977.

Banks

> *Banking establishments are more dangerous than standing armies.*
> —Thomas Jefferson

Financial institutions are another set of agribusiness input firms; the input they provide is capital. As farmers expand their farm production, they have to rely increasingly on nonfamily and nonfarm sources of finance. This development has allowed lending institutions to gain considerable control over the decisions farmers make. As the power of financial institutions has grown, the power of farmers has declined.

Farmers borrow money to finance the expansion of their acreage, to buy larger machinery, and to pay for agricultural services and supplies before their crops are harvested and their farm products are sold. In 1978, the total farm debt rose to $120 billion, $64 billion in real estate debt and $56 billion in nonreal estate debt. Retiring farmers who sold their land to other farmers provided 35 percent of all real estate credit, commercial and savings banks provided 26 percent, and the federal government provided the remaining 39 percent (see Table 7.2 and Figure 7.5).

The Farmers Home Administration (FmHA) illustrates the role of the state in strengthening capital and agribusiness power while perpetuating the myth that family farmers exist and are viable. The FmHA was established to be the lender of last resort, to fill the farm credit needs that remained unmet by private sources. Such loans are inevitably those that are unprofitable to private

TABLE 7.2
Lenders of Farm Credit, 1978

Real Estate Source	Billions of Dollars	Percentage
Sellers of Farms	22.3	35
Insurance Companies	8.7	14
Commercial Banks	7.8	12
Federal Land Banks	21.5	33
Farmers Home Administration	4.0	6
	64.3	100
Nonreal Estate Credit Source		
Commercial Banks	25.7	46
Production Credit Administration	13.5	24
Merchants, Dealers, and Individuals	8.2	15
Others	8.1	15
	55.5	100

Source: USDA, *Agricultural Outlook*, 1979.

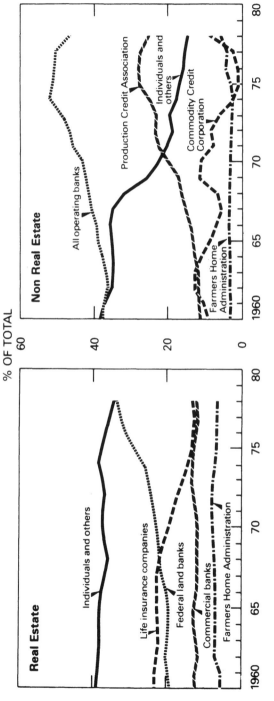

Figure 7.5. Holders of farm real estate and non real estate debt, 1960-1978.

banks. But the FmHA has never had enough funds to meet more than a fraction of farmers' financial needs; its purpose was never to compete with or replace commercial banks. The federal role in providing farm loans has helped to disguise the inability of family farmers to survive in the marketplace and hence to provide a larger population of farm families to be exploited by agribusiness supply and processing firms.

The dependence of family farmers on private banks is most vivid in states with the largest number of remaining family farmers. These same states experienced the highest average increase per acre in the value of farm real estate from 1972 to 1977 (see Figure 7.6). With the value of farm real estate rising, farmers are able to use their land as collateral to borrow more money. Although agricultural loans are a small portion of all loans from U.S. banks, such loans are of considerable significance in these same geographical areas. The Federal Reserve districts with the highest ratio of agricultural loans to total loans in 1974 were Kansas City, Minneapolis, and St. Louis—regions where small-scale farmers are still common. These districts ranged from three to five times the national ratio of 4.8.[13]

The fate of family farmers is intricately tied to the survival of small-town banks. Small rural banks that are heavily involved in farm lending are most vulnerable during a time of falling farm incomes. One-third of all commercial banks were in this precarious category in 1976. These banks accounted for over one-half of all farm loans made by commercial banks. The most vulnerable were the 2,113 banks with 64 percent of their loans in farming. This group represented 15 percent of the nation's banks but accounted for 26 percent of all farm loans outstanding at banks. They held only 2.4 percent or

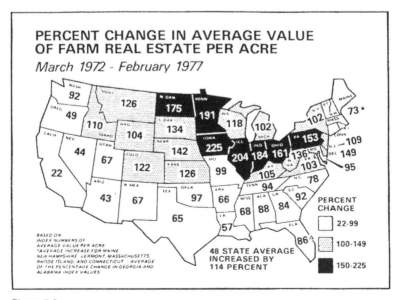

Figure 7.6.

$20 billion of the nation's bank deposits, a figure smaller than the deposits of each of the three largest U.S. banks. All banks have experienced increasing loan/deposit ratios, but this trend is most disturbing for the undiversified smaller banks. Falling farm incomes mean a lower growth in bank deposits and at the same time a greater demand for farm loans. If this inherently risky situation worsens, smaller banks will be in financial trouble, which will lead to their take-over by larger big city banks or to active involvement of the state to shore them up. Small, undiversified banks whose farm loans represent a high percentage of their total loans are thus both exploiters of family farmers and exploitable by larger banks. Under a more concentrated banking system, larger loans would be made available to fewer, large-scale producers, since bigger banks have higher legal loan limits than small rural banks. The managers of large banks place great emphasis on bookkeeping and cash-flow records and therefore favor already heavily capitalized and specialized farming operations.[14]

Financial institutions also exploit farmers through interest rates (see Figure 7.7). The highest interest rates for real estate loans in 1978 were charged by life insurance companies (9.3 percent) and federal land banks (8.2 percent). The lowest rates were charged by former farmers (7.8 percent). Interest rates on nonreal estate farm loans showed a similar pattern: large commercial banks and rural banks charged 9.3 percent in 1978, and production credit associations, reflecting the views of farmers somewhat more directly, charged 8.6 percent.[15]

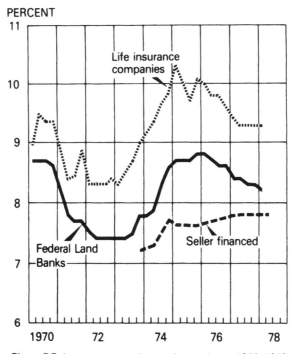

Figure 7.7. Interest rates on farm real estate loans, 1970–1978.

Because all farmers, including family farmers, are now borrowing more money than ever, the significance of family farming is being lost through financial dependency and exploitation. Proponents of agribusiness measure real estate debt "as a percent of value of land and buildings owned and operated" (assets). For all U.S. farms, real estate debt represented 28 percent of all farm assets in 1974. (See Figure 7.8 for the extent of total U.S. farm indebtedness.) But when total debt (real estate and nonreal estate) is expressed as a percentage of the value of production expenses, the figure jumped to 117 percent. Debt-to-asset ratios are poor indicators because most farm assets are not liquid; they are tied up in land, machinery, and buildings, which cannot be used to pay current and on-going expenses. Debt-to-income ratios are a far more realistic measure of indebtedness. While debt on real estate rose 425 percent, debt on equipment rose 380 percent, and debt on operating expenses rose 300 percent between 1960 and 1975, net farm income in the same period rose only 150 percent. On the basis of value of farm product sales, farms with the highest total debt in 1974 as a percentage of net farm income were those with sales of less than $20,000, followed by farms with between $20,000 and $39,999 in sales. Farms with sales below $500,000 to $40,000 had debt/net farm income percentages of 230 to 286.[16] The total U.S. farm debt as a percentage of net farm income was 356 percent in 1974! In other words, farm debts were 3.5 times greater than farm incomes. Nonfarm income helped to reduce this ratio to 2.3. Nonetheless, U.S. farm families owe their capital assets, their managerial talents, and their labor to financial institutions. In short, family farmers can hardly be said to be independent; they are in hock to banks.

The financial attractiveness of farming as an occupation is thus disappearing. As agricultural economist Emanuel Melichar observed: "the current capital spending and land price boom has been debt financed to an extent not experienced since 1920. Thus, some apprehension about the near-term finances of the farming sector appears justified."[17] Peter Divizich, a grape farmer from the Delano-Ducor area of California, reveals the tenuous relations farmers have with banks. Although his borrowing relations with the Bank of America went back almost thirty years, he was faced with foreclosure when he was unable to repay a loan at the end of one crop year. This bank action led to the loss of his ranch, which was valued in excess of $12 million.[18] Young farmers, who are especially caught in the cost-price squeeze, get deeper in debt as their land values increase. As many of them default, they are forced to leave farming. If this involuntary exit continues, family farming has only about twelve years left, according to University of Missouri agricultural economist Harold Breimyer, before it is completely replaced by agribusiness.[19]

Banks with their conservative lending policies prefer to lend to profitable farmers, who are usually large-scale farmers. If family farmers receive credit, they quickly find that they are on a treadmill of ever-increasing debt to finance increasing production to pay this debt. As University of Wisconsin farm records specialist Bob Luening pointed out, "There is a growing tendency for farmers to be perpetually in debt" for the entire length of their farming lives.[20] Traditionally, farmers have abhorred the thought not only of debt in

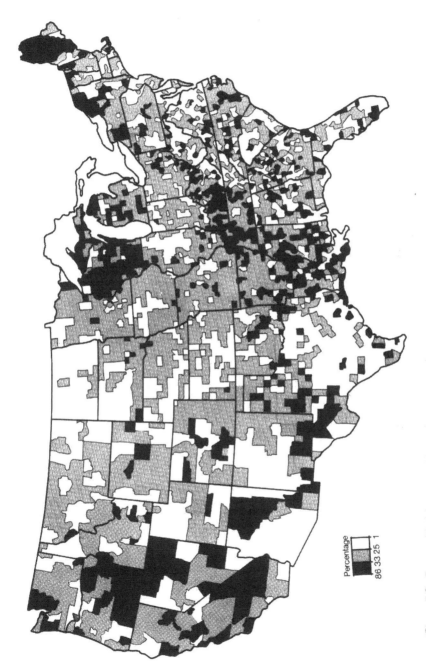

Figure 7.8. Farms with debts secured by real estate as a percentage of all farms, 1974. *Source: 1974 Census of Agriculture.*

general but also of paying interest. Debts allow banks to influence the management decisions of farmers and interest represents the loss of wealth produced by farmers. But within the context of agribusiness, farmers must borrow money, go out of business, or have a subsistence standard of living. Although farmers are borrowing more money, their living standards are falling. Fewer farmers are working with more capital, producing more food and fiber, receiving lower prices, and getting farther in debt! Increasing indebtedness coupled with decreasing net farm incomes places the destiny of farm families in the hands of financial institutions and renders the meaning of the family farm mythical.[21]

Processing and Marketing Oligopsonies

The trend toward greater concentration of economic power is also occurring in the farm processing and marketing sectors. The total number of food manufacturing plants dropped from 42,000 in the early 1950s to fewer than 27,000 in the 1970s. In the flour milling industry alone, the number was halved during the 1950s and 1960s. By the end of the 1960s there were fewer than 540 mills still in business. In 1965, the 31 largest firms had more than 40 percent of the total milling capacity in the United States. Plants processing poultry have declined in number and increased in size. While the number of federally inspected plants was falling 15 percent between 1964 and 1970, the number of large plants processing 30 million or more pounds of poultry a year doubled. These large plants account for about 80 percent of total poultry output. The Federal Trade Commission found that 4 (Campbell, Heinz, Del Monte, and Libby) of the 1,200 canning companies made 80 percent of the canning profits. During the 1970s 6 grain companies bought 90 percent of all U.S. grain and 2 U.S. corporations handled 50 percent of the world's grain shipments. Twenty-four of the 32,500 food manufacturers accounted for 57 percent of the sales.[22]

Fresh vegetable producers sell to food wholesalers, who have doubled their sales per establishment since the early 1950s. Wholesale functions are being integrated with retailing—often, to create field-to-shelf integrated systems. Four out of every five firms in the supermarket industry have their own central warehouses or have other integrated suppliers.[23] Integrated farm production firms face the downward pressure on farm prices just as farmers do, but what the food giants give up in farm prices they take in from other sectors of their operation, such as processing and packaging. Individual farm families are unable to do this. For example, farmers received only 3.5 cents from a 30-cent can of Del Monte tomatoes sold in 1973.[24]

Agribusiness must create consumer demand for food items to unload its products. To get consumers to buy food from a particular company, brand differentiation and brand loyalties are created through massive advertisement campaigns and packaging. Brand names replace generic foods: Sunkist oranges substitute for oranges, Green Giant sweet peas for peas, General Mills' Wheaties for grains. With the growing emphasis on merchandising food, marketing firms will gain ascendancy over farm-supply firms, concentrating economic power even more unevenly within agribusiness itself.

All farmers, especially full-time, small-scale producers, are squeezed by agribusiness firms from two directions, as both buyers and sellers. The concentrated farm-supply industry determines the prices farmers *pay* for essential farm inputs (oligopoly), and the concentrated food-processing industry determines the prices farmers *receive* for their products (oligopsony). In 1970 agribusiness input and processing firms received 88 percent of the market value of agricultural products in the form of production expenses. By 1977 agribusiness firms took 92 percent of the market value of farm products in production expenses.[25] Although farmers directly took the risks of uncertain weather and disease conditions and of the market, they received only $8 billion in net farm income from the total market value of $96 billion.

The exploitation of farm labor has been a persistent pattern. Although prices farmers paid for inputs went up 52 percent between 1951 and 1971, the prices they received for their products increased only 8 percent. During the 1970s, farmers were paying more for inputs than they were receiving for their products in every year except two. Farmers received substantially higher prices during 1973 and 1974 because of strong foreign demand. Consequently, net income from farming after inventory adjustments (measured in constant 1967 dollars) was highest from 1972 to 1975. But net farm incomes in 1976 were below those of 1970 (see Figure 7.9)! Income from farming is so inadequate that farm families, on the average, earned nearly 60 percent of their income from off-farm sources in 1977. The average income of farm families was $11,600 from off-farm sources, while the average net income from farming was only $7,500. Farm families are distinctive for their multiple employment status among U.S. workers *and* for having a lower per capita income than other workers. Since the mid-1960s, disposable personal income per capita for the farm population has averaged, at the very best, 80 percent of that for the nonfarm population (see Figure 7.10).[26]

Agribusiness can effectively squeeze farmers even more by tying farm production directly to its needs. Huge seed and feed companies; chemical and fertilizer producers; farm machinery manufacturers; processors, canners, and packagers; and marketers and distributors are increasingly moving into production either directly by operating farmland or indirectly by signing production and/or marketing contracts with "independent" farmers. Through vertical integration agribusiness is eliminating farmers, and through contract farming agribusiness further erodes the decision-making powers of the remaining farmers.

Horizontal and Vertical Integration

Firms that operate similar kinds of businesses are *horizontally integrated*. Monfort of Colorado is a good example. The company operates two 400-acre, 100,000-head feed lots in Greeley and Gilcrest, Colorado, with an annual production of over 400,000 head. Horizontally integrated firms apply their skills in the same line of business in different places. Many of these companies are also *vertically integrated*, operating different levels of the same business

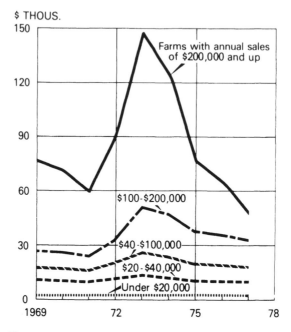

Figure 7.9. Net income per farm by sales classes, 1969–1977.

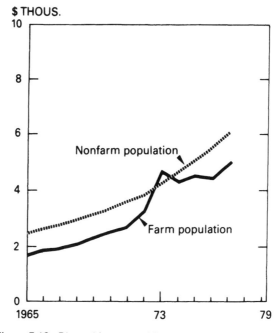

Figure 7.10. Disposable personal income per capita of farm and nonfarm population, 1965–1977.

to control all aspects of production and/or marketing. Vertical integration allows companies to determine the quantity, quality, and price of their inputs. Also, profits from each subsidiary are retained by the parent corporation, giving it greater economic clout. These firms have far greater flexibility in where they make their overall profits than single-business firms. Unlike family farmers, they can afford to actually lose money in one sector, for example, farming, because they can make up the profit in another sector, such as marketing. Vertical integration depends on controlling the food production system from beginning to end, irrespective of whether this control rests on direct ownership, leasing of resources, or legal contracts with others who provide or use resources. The Monfort feed lots (see Figure 7.11) are vertically integrated. Monfort contracts with local farmers for corn ensilage and green chopped alfalfa, providing seed, technical assistance, and harvesting for the majority of these crops. The company also owns and operates elevators at Cozad, Nebraska, and Goodland, Kansas. The corn from these elevators is shuttled to the feed lots by ninety company-leased railroad cars and supplementary transportation of feed is provided by a fleet of company-owned grain trucks. Finally, the company owns the Monfort Packing and Monfort Provision companies, which under contract provide Hilton Hotel restaurants with prime cuts of meat and Mr. Steak Restaurants with less-than-prime cuts.[27]

Tenneco, the thirtieth largest U.S. industrial corporation, is probably the best-known vertically integrated food corporation. In 1967 it acquired the Kern County Land Company, California's third largest landowner. Three years later it took control of Heggblade-Marguleas, the nation's largest marketer of fresh fruits and vegetables. Tenneco also owns J. I. Case Company, which manufactures, among other things, farm machinery, and the Packaging Corporation of America, which manufactures food containers. Sun Giant Supermarkets, another subsidiary, retails Tenneco's food.[28] In short, "Tenneco plows its own land, fertilized and sprayed with chemicals from its own chemical division, using its own tractors fueled with gas and oil from its own oil wells and refineries. The food is processed, packaged, and distributed by Tenneco subsidiaries."[29]

Tenneco is only one of many integrated agribusiness firms. Arizona-Colorado Land and Cattle Company is a less well known one, but it provides an excellent example of the profitability of vertical integration. By 1973, the losses of the 1960s were replaced by a compound growth rate of 37 percent in revenue, 53 percent in operating income, and 94 percent in common equity. It manages large tracts of agricultural land for immediate revenue from livestock and crop operations, which support the carrying costs of the land until it can be further developed for industrial parks and residential subdivisions. On 400,000 acres of deeded land and 700,000 acres of leased land in Arizona, Nevada, Colorado, New Mexico, and Florida, the company raises calves and feed cattle for its own or other feed lots and fattens cattle for its own or other packing plants. Each stage of the operations is planned to offset any major pricing problems at another stage. The company has also acquired several consulting and engineering firms with expertise in land, water,

Figure 7.11. Agribusiness feedlots. (Top) One of the 120 employees watches a television monitor to see when a truck is in position to receive its load of computer-calculated feed ingredients at the Monfort feedlot. She releases the ingredients into the truck and tells the operator by radio which pen gets the feed. Michael Lawton, USDA photo. (Bottom) Trucks mix the computer-calculated feed on way to the 265 separate feeding pens, which hold about 100,000 head of cattle at any one time. Michael Lawton, USDA photo. According to Herman Koenig, director of Michigan State University's Ecosystem Design and Management Program, feedlots of 200 to 300 head of cattle are the optimum size for efficient energy and resource use.

and mineral developments and two manufacturers of farm equipment "useful during farm industry booms."[30]

Agribusiness has been very selective in the crops and livestocks that have been vertically integrated (see Table 7.3). In 1970 the major crops were sugarcane, citrus fruits, vegetables for fresh market, potatoes, other fruits and

TABLE 7.3
Production Contracts and Vertical Integration of Crop and Livestock Products, 1970

Item	Production/Marketing Contracts	Vertical Integration
	Percent	Percent
Feed Grains	0.1	0.5
Hay and forage	0.3	--
Foodgrains	2.0	0.5
Vegetables for Fresh Market	21.0	30.0
Vegetables for Processing	85.0	10.0
Dry Beans and Peas	1.0	1.0
Potatoes	45.0	25.0
Citrus Fruits	55.0	30.0
Other Fruits and Nuts	20.0	20.0
Sugar Beets	98.0	2.0
Sugarcane	40.0	60.0
Other Sugar Crops	5.0	2.0
Cotton	11.0	1.0
Tobacco	2.0	2.0
Oil Bearing Crops	1.0	0.4
Seed Crops	80.0	0.5
Miscellaneous Crops	5.0	1.0
Total Crops	9.5	4.8
Feed Cattle	18.0	4.0
Sheep and Lambs	7.0	3.0
Hogs	1.0	1.0
Fluid Grade Milk	95.0	3.0
Manufacturing Grade Milk	25.0	1.0
Eggs	20.0	20.0
Broilers	90.0	7.0
Turkeys	42.0	12.0
Miscellaneous	3.0	1.0
Total Livestock Products	31.4	4.8

Source: R. L. Mighell and W. S. Hoffnagle, Contract Production and Vertical Integration in Farming, 1960 and 1970, USDA, Economic Research Service Report No. 479 (Washington, D.C.: Government Printing Office, 1970).

nuts, and vegetables for processing. The major animal products were eggs and turkeys. In total, about 5 percent of all crops and livestock commodities were vertically integrated.[31]

Vertical integration allows agribusiness to extract profits at every stage of the food production process, an advantage that family farmers do not share. In the past twenty-five years farmers have received 16 percent more for their crops while the nonfarm costs in food have gone up more than 80 percent.[32] Whether integrated firms expand into farming to assure themselves adequate supplies of raw materials, or to reduce their taxes on nonfarm income, or to increase overall company profits, they adversely affect the survival of family farmers. Agribusiness displaces family farmers with large-scale producers who are less technically efficient but more profitable and more tightly integrated into agribusiness.

Types of Corporate Farms

The corporate invasion of American agriculture by nonfarm interests is real. It is leaving behind wasted towns, deserted communities, depleted resources, empty institutions and people without hope and without a future. The invasion is still in the beginning stage. Some people see this trend as inevitable—that is, cannot be stopped. Not only can it be stopped, it must be stopped!
      —Tony T. DeChant, president of the National Farmers Union

The U.S. census classifies farms into three legal organizational types: sole proprietorship, partnership, and corporation (see Table 7.4). Sole proprietorships accounted for 89.5 percent of all farms in 1974. These farms are thought to be associated with family farm operations, yet sole proprietorship in no way ensures that farms are operated by someone living on the farms and working the land with other members of the family. Similarly, partnerships— representing 8.6 percent of farms—are thought to be parent and child or several family member operations. Again, partnerships can represent large-scale operations and may even have nonfarm corporations as one partner.

For many Americans, corporations, which accounted for 1.7 percent of all farms, represent large nonfarm companies invading agriculture and displacing family farmers. The corporate form of ownership, however, is not the problem; in fact, the corporate structure has several advantages for farmers of every size:

1. ease of transferring interests in property by transferring shares of stock for estate planning
2. planning management and ownership succession to make continuation of the business easier after the deaths of the original owners
3. avoidance of full-owner liability for obligations of the business through shareholders' limited liability
4. opportunities for income tax savings

Figure 7.12. Agribusiness vertical integration. White Hen Poultry Farms, a subsidiary of Jewel grocery stores, produces 100,000 eggs each week outside Chicago. This 150-acre "farm" has sixteen automated chicken houses containing 50,000 hens per building.

In a legal sense, corporate farms are any farms that are incorporated under their respective state laws. But in a structural sense, corporate farms represent three very different classes: family-farm corporations; family-owned, larger-than-family corporation farms; and nonfamily corporation farms. Most of the 1974 census "family corporations" were not family farms because they employed more than 1.5 man-years of nonfamily labor. Nonfamily corporation farms include large corporations whose stocks are frequently listed on stock exchanges and publicly traded. Among them are Tenneco, Beatrice Foods, General Foods, Coca-Cola, United Brands, Bank of America, Southern Pacific, and Ralston Purina.[33] Such corporations are typically involved in farm production, agricultural supply, and/or processing activities.

Nonfamily corporation farmers have several additional incentives to incorporate. Corporations generally can secure better-trained management than sole proprietorships or even partnerships and thus capitalize more thoroughly on technological changes. The corporate structure also provides easier access to external funds because of the continuity of management and of the company (which does not depend on the survival of one individual). Large corporations can decrease risk and uncertainty by controlling inputs, marketing, and, to some extent, prices. The capacity of giant corporations to grow,

TABLE 7.4
Legal Structure by Farm Size, 1974

Legal Structure	Average		Large-Scale Farms				Medium-Scale Farms		Small-Scale Farms			
			Largest: Sales of $200,000+		Large: Sales of $100,000-199,999		Sales of $40,000-99,999		Small: Sales of $10,000-39,999		Part-time & Subsistence Sales of Less Than $10,000	
	Farms	Acres	Farms	Acres	Farms	Acres	Farms	Acres	Farms	Acres	Farms	Acres
Corporations	1.7	10.7	20.6	35.0	5.2	15.0	1.7	7.4	0.7	2.6	0.4	1.7
Partnerships	8.6	13.8	21.4	22.2	16.7	19.0	11.4	14.0	8.0	9.8	5.0	6.9
Individual Owners	89.5	75.4	57.9	42.8	78.1	66.0	36.8	78.6	91.2	87.6	94.5	91.4
	100.0	100.0	100.0	100.0	100.0	100.0	100.0	100.0	100.0	100.0	100.0	100.0

These figures include farms with sales of $2,500 and over.

Source: General Accounting Office's analysis of 1974 Census of Agriculture data found in Ed Schaefer, Changing Character and Structure of American Agriculture: An Overview (Washington, D.C.: General Accounting Office, 1978).

despite undesirable social consequences and lack of increased production efficiencies, is the crucial difference between large-scale corporate farms and individual family farms.

The corporate structure tends to separate labor, management, and capital. In corporate factories in the fields, labor is hired and division of tasks is imposed. Farmers become tractor drivers or mechanics. Corporate agriculture radically changes decision making; the strategic, long-range planning and financial decisions are made in distant, corporate headquarters. Managers are responsible for everyday decisions based on a game plan issued from above. The large capital needs of agricultural corporations and of agricultural divisions of industrial conglomerates are served by the national capital market, operated by other large corporations and banks.

Many corporation farms are structurally family farms. According to the *1974 Census of Agriculture*, 76.5 percent of all corporate farms were owned by family members; 18 percent were owned by independent nonfamily corporations; and 5.5 percent were held by corporations that owned or controlled one or more other corporations. Only 3.6 percent of all corporate farms were each owned by eleven or more shareholders, and only 33 percent were publicly owned and traded on the stock market. In other words, corporate farms are generally owned by a few related or unrelated persons. Yet family ownership and control of farming corporations are not synonymous with family farms. Cargill, the largest grain merchant in the world, is a family-owned corporation; its stocks are not traded and consequently little is known about it. [34]

Corporate farms are legal entities; family farms are a specific structural form of farming. The legal status of farms is not particularly important in understanding the structure of U.S. agriculture, but since corporate farms, whether family controlled or not, are frequently large-scale operations, they become significant. Census data from 1974 indicate that 28,442 corporate farms held 1.3 percent of all land in farms and produced 17.9 percent of the total value of farm products. The average corporate farm consisted of 3,380 acres, had land and buildings valued at $866,000, and produced $515,000 in farm products. Although the vast majority of corporate farms were owned by a small number of family members (usually less than 5), the size of these farms places them in the larger-than-family and industrial farm categories.

Although the legal status of incorporation has advantages for farmers in all size categories, corporations are most prevalent in the larger farm product sales classes and account for a higher proportion of farm sales. Almost all corporate farms were found in the $100,000 and over sales categories in 1974. In the largest category ($500,000 and over), corporate farms accounted for 40 percent of all farms and 50 percent of all acres owned. A General Accounting Office study of unpublished 1974 census data found that of the 4,040 farms with sales of at least $1 million, 2,330 were corporations, 700 were partnerships, and 940 were sole proprietorships or individuals. In other words, 8 percent of all farm corporations had sales of $1 million and over, while only 0.5 percent of partnerships and 0.1 percent of individual farms were that large. [35]

Nonfamily corporate farm operations involved high risk, high return capital-intensive specialized crops and livestock production systems associated with integrated agribusiness. Corporations tend to concentrate on fresh and processed vegetables, cotton, cattle, and poultry. Corporation farms are relatively few in low-risk products that have not yet been integrated with processing firms: hay crops, wheat, and beef range operations. Nonfamily corporations decrease their costs by operating in areas with uniform soil and climatic conditions, hence corporate farms are particularly concentrated in the irrigated West and humid South[36] (see Table 7.5 and Figure 7.13). The profitability of such a tactic is demonstrated in the mid-Columbia Basin of Washington and Oregon (see Figure 7.14). Since 1972, corporations such as Boeing Aerospace Co., Utah-Idaho Sugar Company, I & U International

TABLE 7.5
States With Major Concentrations of Corporation Farms

State	Corporation Farms as a Percentage of All Incorporated Farms[a]	
	Farms	Land in Farms
Hawaii	22.5	64.4
Arizona	13.2	34.7
California	9.9	24.6
Connecticut	9.8	24.4
Texas	9.1	23.3
Illinois	8.7	22.3
Nevada	8.3	33.0
Massachusetts	8.1	19.5
Florida	7.8	36.2
Arkansas	7.6	7.4
Georgia	7.3	7.5
Delaware	7.1	7.9
New Mexico	6.8	21.6
Oklahoma	6.5	20.4
U.S.	5.5	15.4

[a]According to the census, incorporated farms consist of three types: "independent, family, and parent." The latter are controlled by other corporations--only these are shown here.

Source: 1974 Census of Agriculture.

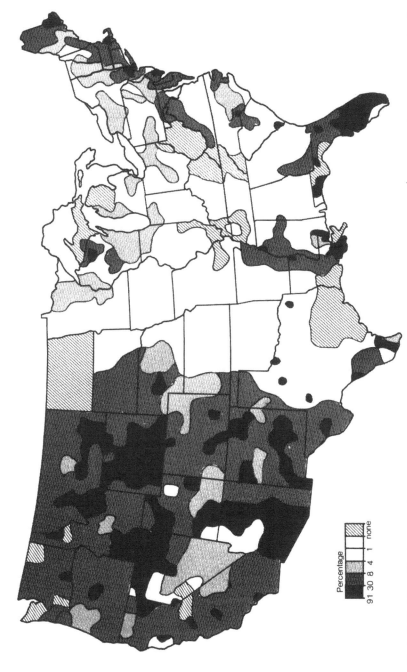

Figure 7.13. Generalized corporation acreage as a percentage of total farm acreage, 1974. (North Dakota law prohibits farm corporations.) *Source: 1974 Census of Agriculture.*

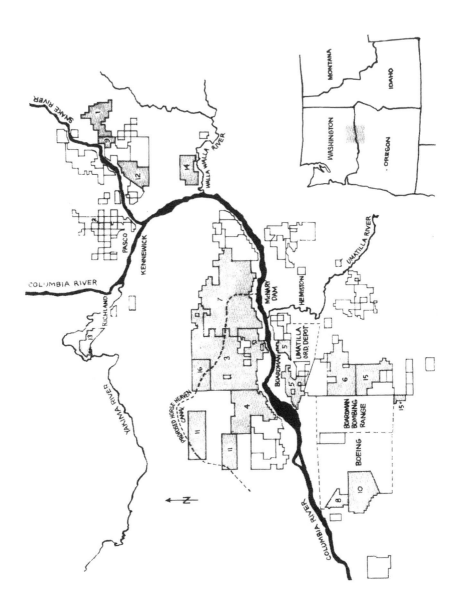

Irrigated Land Ownership in the Mid-Columbia Basin

No.	Property	Location	Owner	Acres
1	K2H	Walla Walla. Wash.	U & I Sugar Co	11,500
2	Burlington Northern	Pasco, Wash.	Burlington Northern Inc.	3,500
3	Prior Land Co.	Horse Heaven Hills. Wash	U & I Sugar Co	20,400
4	Columbia River Farms	Horse Heaven Hills. Wash	AMFAC. Inc	8,000
5	Oregon Potato Inc.	Irrigon, Ore.	C Brewer Ltd. (sub. of IU International)	8,650
6	Sabre Farms	Boardman, Ore.	Idaho & Washington businessmen	13,900
7	Horse Heaven Farms (proposed)	Horse Heaven Hills. Wash.	U & I Sugar Co.	16,500
8	Boeing Agri-Industrial Company	Boardman, Ore.	Boeing Aerospace Co.	8,500
9	Snake River Farms	Walla Walla. Wash.	Partners P.J. Taggares. E. J. Spiegal. A. Wolcott. Universal Lands	3,000
10	SimTag	Boardman, Ore.	P.J. Taggares, J.R. Simplot	12,500
11	Quadrant (proposed)	Horse Heaven Hills. Wash.	Weyerhaeuser Corp.	31,000
12	Universal Lands Co.	Walla Walla. Wash.	n.a.	6,000
13	Lewis % Clark Angus	Richland, Wash.	n.a.	5,000 (proposed) 7,000
14	Two Rivers	Walla Walla. Wash.	n.a.	6,700
15	Larry Lindsay Farms	Boardman, Ore.	Larry Lindsay	4,600
16	Munn (proposed)	Horse Heaven Hills. Wash.	Munn Group	10,000

Figure 7.14. Large private and multinational corporation landholdings in the mid-Columbia Basin. Source: The AgBiz Tiller, No. 1 (August 1976), pp. 2–3. San Francisco Study Center, P.O. Box 5646, San Francisco, California 94101.

Corporation, AMFAC, Burlington Northern, J. R. Simplot Company, P. J. Taggares Company, and U.S. Tobacco Company have established large vertically integrated food production, processing, and manufacturing operations. The key to this corporate expansion is pivotal irrigation: water is pumped from the Columbia River to the center of round fields a half-mile in diameter. Giant arms of six-inch pipe, a quarter of a mile long, pivot around the centers of the fields every twelve hours. The arms are supported ten feet off the ground by ten towers; each tower rolls along on two rubber-tired wheels driven by electric motors. Sprinklers along the length of the pipes shower the ground continually throughout the long growing season, dropping the equivalent of sixty inches of rain annually on land that naturally only gets seven inches per year (see Figure 7.15). About 170,000 acres of desert have already been irrigated in just five counties lying on either side of the Columbia. The reclamation projects require huge amounts of capital that only large corporations can secure—$100 million since 1970. But the payoffs are immense. For example, in 1973, 120,000 unirrigated acres in Morrow County, Oregon, yielded $8.2 million worth of wheat, yet a mere 6,000 irrigated acres a few miles north produced $9.2 million worth of potatoes.[37]

J. Phil Campbell, the undersecretary of the U.S. Department of Agriculture (USDA), concluded in 1972 that the involvement of nonfamily types of corporations in farming was extremely low. Richard Rodefeld, in his critique of the USDA data and conclusions, has pointed out that the structural characteristics of corporate farms include large acreage and/or sales, absentee owners, hired managers, and hired field workers. "If this structural definition of a 'corporate' farm is accepted, then it is clear the USDA definition of a 'corporate' farm not only does not include all 'corporate' farms: it probably does not even include 'most' corporate farms."[38] Farms with structural characteristics of corporate farming need not be legally incorporated. From research on Wisconsin, Rodefeld reported that 75 percent of all structurally defined *and* legally incorporated farms were owned by an individual, family, or small group of unrelated individuals. These "corporate" farms were excluded from the USDA's "corporate" farm category because they were operated by "individuals." The exclusion of structural corporate farms in Wisconsin was probably even greater than 75 percent because no assessment was possible on the number of such farms unincorporated.[39]

Rodefeld also analyzed the USDA's 1968 publications on corporations in farming and found that the USDA underestimated corporate-operated acres (structurally defined) by 46 percent, acres rented by 298 percent, and cattle fed by 46 percent. USDA data from this survey even conflicted with census data: the 1969 census found 61 percent more farming corporations than the USDA survey![40] In Minnesota and Iowa, independent studies indicate that the underestimation of the number of corporate farms was over 50 percent.[41]

Another inadequacy of the USDA's enumeration of corporate farms is revealed by the Internal Revenue Service. The IRS found a 22 percent increase in corporate farms in the mid-1960s because the USDA survey had omitted corporations that rented out all the land they owned.[42] Some of the USDA's classifications are truly absurd. Agribusiness corporations that contract with

Figure 7.15. Many sandy and/or dry regions of the United States are being turned into profitable crop land through massive capital investments, which only large-scale farm operators can afford. Pictured is an example of center-pivot irrigation in the Mississippi River Valley, Minnesota.

farmers to produce crops or livestock products are not considered corporate farmers by the USDA. Instead, these processors are considered "subsistence farmers" since they technically consume all they produce.

The absence of a structural definition of corporate farming renders the USDA statistics meaningless. Senator Gaylord Nelson of Wisconsin, who conducted hearings on corporation farming for the U.S. Senate Antitrust and Monopoly Subcommittee, summarized the visibility of corporate farms well. He said that corporate farming "can best be compared to a giant iceberg. Only about 10 percent of it is above the surface and open to public view. The remainder is hidden in a tangle of corporate connections and deceptive arrangements."[43]

The relatively small number of corporate farms has increased rapidly. Indeed, over half of all existing U.S. farming corporations have been established since 1960. From 1969 to 1974 their numbers increased by 25 percent. From 1969 to 1974, the number of reported corporate farms in Minnesota doubled to 700.[44] Given a history of rising farmland values over the past 35 years, with land being a good hedge against high rates of inflation in the 1970s, and with a tax policy favoring corporate types of investment for high-tax-bracket individuals, the recent expansion of corporation farming is not surprising and predicts future expansion at an accelerated pace.[45]

Large-Scale Farms

The lives of farm families and rural residents are affected not by the legal status of certain farms but rather by the kind and size of farms. Incorporated family farms are essentially indistinguishable from nonincorporated family farms. But large-scale farms, whether incorporated or not, play a major role in the life of rural communities. With food firms trying to minimize their costs and inconvenience by buying and assembling farm products in large volume, large-scale producers are distinguishable from small-scale farmers in four ways.[46] First, large-scale farmers seek profits comparable to other commercial firms in manufacturing and service industries. Agricultural economists Charles Moore and Gerald Dean expressed it this way: "Family farms will accept low rates of return on their resources because of nonprofit goals, but industrialized firms are likely to be guided primarily by profits."[47] Second, these farmers are at such a large scale of production that the number and complexity of management decisions are well beyond the ability of a single owner-operator. Industrialization is thus synonymous with the specialization and decentralization of management. Third, large-scale farmers have greater access to capital to purchase capital-intensive technology and to expand their productive land base. Fourth, managerial complexity and heavier capital dependency require that large-scale farmers reduce risks of income variation. Such reduction is achieved by concentrating largely on high-value perishable goods that are tightly integrated into other controllable sectors of agribusiness, for example, canneries and slaughterhouses.

The ability of large-scale farmers to increase capitalization of land and nonland resources allows these farmers to produce over one-third of the total sales on less than one-seventh of the land. The largest 2.1 percent of farmers had nearly 37 percent of sales but only 14 percent of land[48] (see Table 7.6). The growth of factory-type production techniques (confinement feeding in the beef, poultry, and egg industry in addition to greenhouse production) allows relatively small-acreage farms to produce a disproportionate volume of sales. In 1974, farms with $100,000 and over in sales had about 70 percent of all hired and contracted labor expenditures. According to Table 7.7, farm income and debt are concentrated on the largest farms. In 1974, average farm income for the largest farms was nearly 600 times greater than the average farm income of the smallest farms. The largest farms also have twice the indebtedness of the smallest ones. The organizational traits of large-scale farms make these farms the logical extension of agricultural input and processing firms. *Large-scale farms represent the production side of agribusiness.*

Contract Farming

Agricultural input and processing firms tie farm production to their needs by signing contracts with large-scale *and* small-scale producers. Contract farming is particularly negative for family farmers because it erodes their independence. "Independent" farmers sign contracts with specific companies to deliver a certain quantity and quality of farm products at a set price.[49] Under

TABLE 7.6
General Characteristics of Farms, 1974

Characteristics	Large-Scale Farms — Largest: Sales of $200,000+	Large-Scale Farms — Large: Sales of $100,000-199,999	Medium-Scale Farms — Sales of $40,000-99,999	Small-Scale Farms — Small: Sales of $10,000-39,999	Small-Scale Farms — Part-time & Subsistence: Sales of Less Than $10,000
General					
Number of Farms	51,446	101,153	324,310	631,782	1,203,084
Percent of Farms	2.1	4.5	14.0	27.3	52.2
Average Farm Size (acres)	2,826	1,299	761	416	203
Percent of Land	14.1	12.8	24.1	25.7	23.3
Value of Farm Sales ($)	581,996	136,012	61,890	21,969	5,321
Percent of Farm Sales	36.8	17.0	24.7	16.9	4.7
Farm Type by Percent					
Cash Grain Farms	2.2	6.55	22.19	40.59	28.44
Livestock Farms, except dairy, poultry, and animal specialty	2.6	4.5	13.2	31.2	48.6
Dairy Farms	2.1	6.0	31.8	50.6	9.6
Vegetable and Fruit Farms	6.1	6.3	15.6	37.1	35.0
General Purpose, primarily crop and livestock	2.8	5.3	18.8	43.8	29.3
Farm Labor					
Hired					
Percent of Farms	6.47	10.29	25.57	34.29	23.37
Percent of Dollars	56.17	15.54	16.57	9.11	2.60
Contracted					
Percent of Farms	8.67	10.09	22.26	32.81	26.17
Percent of Dollars	57.08	12.22	14.74	11.13	4.84

Source: See Table 7.4.

TABLE 7.7
Farm Liabilities and Income by Farm Size, 1974

Liabilities and Income	Large-Scale Farms		Medium-Scale Farms	Small-Scale Farms	
	Largest: Sales of $200,000+	Large: Sales of $100,000-199,999	Sales of $40,000-99,999	Small: Sales of $10,000-39,999	Part-time & Subsistence: Sales of Less Than $10,000
Farm Liabilities					
Percent of Farms in Debt	59.0	57.4	51.4	39.3	29.3
Ave. Loan Size of those in Debt ($)	278,512	95,825	54,566	30,234	17,868
Farm Income					
Value of Farm Sales ($)	585,692	136,012	61,890	21,696	5,321
Production Expenses ($)	475,446	94,728	41,438	14,822	5,129
Production Expenses as Percent of Total Farm Sales	81.1	69.6	67.0	68.3	96.4
Ave. Farm Income per Farm ($)	110,246	41,284	20,452	6,874	192
Other Farm-Related Income	122,278	45,208	22,883	9,148	1,523
Custom Work	33.6%	46.0%	48.4%	43.7%	34.8%
Direct Gov't Farm Payments	17.5%	18.6%	20.6%	20.1%	14.6%
Family Off-Farm Income ($)	13,577	8,047	6,713	8,404	10,665
Average Total Income per Farm Family ($)	135,855	53,255	29,596	17,552	12,188
Average Income from Farming	90%	85%	77%	52%	13%

Source: See Table 7.4.

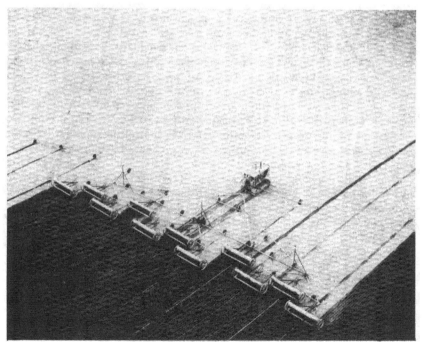

Figure 7.16. Large-scale farming. (Top) Mechanization and automation of rice threshing, 1927. C. L. Snider, Caterpillar Tractor Company and USDA photo. (Bottom) Planting wheat, 1957. Roy Clark, USDA Photo.

contracts, farmers retain ownership of land but lose the management control of their farm operations. By being paid a certain amount per unit of product, contract farmers have almost become wage laborers. They have a transitional status between family farmers and agricultural workers, yet they often continue to bear the risks of farming, paying corporations the difference between the products contracted for and the products delivered.

Not surprisingly, the contracts are not equally binding on both parties. Although farmers must deliver the product to be paid, the processors are liable only for a final price determined by their own graders. As Ruben Reyes, weightmaster for ten years with Libby, McNeill and Libby, explained, "The plant itself purchases the product and they do whatever they want in regards to grading it and paying the farmers. . . . the company I worked for was ripping off the farmers for at least $50,000–$60,000 a year in their deliveries on spinach, altering the grades."[50] Once contracts are signed, processing companies usually make all technical and market decisions about planting and harvesting and provide the necessary farm supplies—fertilizers, pesticides, and harvest machinery.[51] When appropriate, the processing company also lends farmers money to meet the company's specifications in building construction (as in the broiler industry) or to purchase seed and certain kinds of machinery (as in the canning industry). Farmers are thus not only contracted to but also indebted to the processors.

The attractiveness of contracts for agribusiness is that farmers, nominally still their own bosses, have effectively been reduced to employee status. As they are not actual employees, companies can avoid laws that cover industrial workers: minimum wages, workers' compensation, social security, paid holidays, sick leave, and health insurance. *Contracts give agribusiness the advantage of treating farmers as employees without the responsibilities of paying them as employees.* Furthermore, their pseudoindependent status motivates contract farmers to work harder than overtly hired employees. Corporations also escape paying property taxes on land yet control the products from the land as if they owned it. For farmers who experience fluctuating prices, contracts provide minimum and stable incomes. Nevertheless, the disadvantages of contracts for family farmers are that: (1) farmers lose the possibility of higher incomes in the open market; (2) they lose the ability to make short-term managerial decisions and, when specialized buildings are involved, even long-term decisions; and (3) often, their indebtedness increases and is linked to processing firms. In addition, the production from farms under contract can be used by corporations to force down prices of produce from other farmers.

So far, production contracts are important only in a few commodities. In 1970, the major crops involved were sugar beets, seed crops, vegetables for processing, citrus fruits, potatoes, and sugarcane. The major types of livestock products were fluid-grade milk, broilers, turkeys, manufacturing-grade milk, eggs, and feed cattle. In total, 10 percent of all crops and 31 percent of livestock items were contracted, an average of 20 percent. By 1974, 8 percent of all farms were under contract and these farms produced 28 percent of the total $81 billion of agricultural products (see Figures 7.17 and 7.18). In

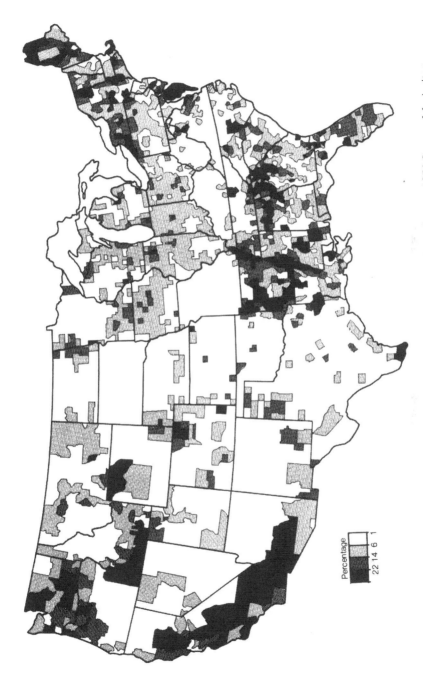

Figure 7.17.   Farms with contracts as a percentage of all commercial farms, 1974. *Source: 1974 Census of Agriculture.*

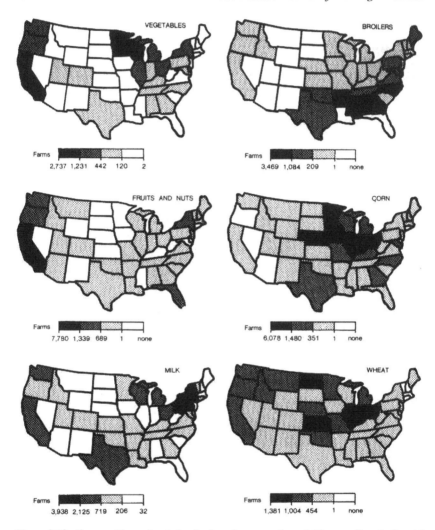

Figure 7.18. Farms with contracts for fresh and processed vegetables; poultry; fruits and nuts; feed and seed corn; milk; and wheat. *Source: 1974 Census of Agriculture.*

1970, the American Agricultural Marketing Association estimated that 50 percent of the U.S. food supply will be produced under corporate contracts by 1980 and 75 percent of the food supply by 1985.[52]

Farmers' reactions to contracts are well illustrated by the dairy industry— characterized by market contracts—and the broiler industry—having both production and marketing contracts. Most dairy farmers, including small producers, consider the methods used to contract for the sale of fluid milk (95 percent of all milk sold) satisfactory. Dairy farmers have a more effective voice and greater input in setting contract terms, but broiler farmers under contract suffer most from the loss of decision-making independence and economic exploitation.[53] In part, the difference results from dairy-producer cooperatives controlling the contractual arrangements for milk while large feed companies, or other nonfarm firms, control the contracting for broilers.

Contracts in the broiler industry have eroded the independence of family farmers to such an extent that even a conservative agricultural economist such as Don Paarlberg admitted, "The farmer who feeds and cares for broilers, formerly a farmer in the tradition of family farming, making his own decisions and risking his own capital, becomes essentially a hired laborer or a piece worker."[54] Reduced to hired labor, poultry producers are earning less by producing more. For example, between 1951 and 1961, California poultry raisers reduced the cost of raising broiler chickens by eight cents a pound. During the same period, the price was lowered eighteen cents a pound—a loss of ten cents a pound in excess of the reduction of cost.[55]

Seasonal tomato farm workers in northeast Ohio have recognized the employee status of the growers for whom they work. When 2,000 to 5,000 of them struck in August 1978, they demanded the right to participate in annual negotiations between growers and canners. Because growers are not actually independent business people but rather are effectively hired employees of the canneries, wages farm workers receive depend on the amount canneries pay farmers.[56]

Yet free-market agricultural economists persist in maintaining that "with respect to his independence as an entrepreneur, it makes no difference whether [the farmer] negotiates prices for specified products in advance or does so in an open market after they are produced."[57] The only conclusion to be drawn from this statement is that the "open market" is already so tightly controlled by the same agribusiness firms as those who sign contracts with farmers that the independence of family farmers has already been severely compromised by agribusiness. In the end, farmers, with or without production and marketing contracts, are at the mercy of agribusiness firms. In many types of farming no open market exists. In the vegetable processing industry, processors themselves produce 10 percent of the total crops and farmers contracted to canneries produce another 85 percent. The National Farmers Organization has identified the heart of the issue: "Farmers sign contracts with integrators because they are hardpressed."[58] Given the high costs of supplies, the low prices farmers receive, and the monopolization of the farmers' market, the choice is often between signing with a corporation or going out of business.[59]

Agricultural processing firms actively encourage mechanization of agriculture, displacement of farm workers, and increased indebtedness of farmers. In northern Ohio, Campbell no longer accepts hand-picked tomatoes. Only farmers who have bought or are buying tomato harvesters receive contracts. One year Campbell reduced its per-ton price by seven dollars from the preceding year to further induce farmers to switch to machines. With a $40,000 harvester, farmers need at least forty acres in tomatoes to make the purchase of the machines worthwhile. Contract farming, then, directly feeds the trend toward more capitalized farms.[60]

Figure 7.19. Contract farming in 1924: marketing strawberries in Chadburn, North Carolina. USDA photo.

Figure 7.20. Henry Heinz, founder of the international food processing company, H. J. Heinz Company, talks to field workers harvesting crops for his "factory-kitchens" in the 1910s. H. J. Heinz Company and USDA photo. In 1909 a field agent for the United States Immigration Commission reported, "[T]he morality, health, and general well-being of these workers seem not to appeal at all to the grower. Although he may speak with disgust of their filth and low standards, it never occurs to him to ameliorate their condition. . . . Labor to many of these men is merely a commodity. The human element rarely enters. It is not likely that any great change will come except as the workers themselves demand it."

The net effect of farm input oligopolies, market oligopsonies, and contract farming is that farmers have become largely employees, frequently being told what to produce, in what quantities and qualities, at what times, and for what prices. Power rests essentially with those who control farm supplies, agricultural markets, and credit. The remaining family farmers have been reduced to assembly-line cogs in the integrated food industry, but most farmers persist in thinking—even defending the idea—that family farms exist and can survive under the capitalist mode of production. Ironically, the more family farms are destroyed, the more tenaciously farmers, politicians, and the public praise this institution. The family-farm myth helps obfuscate reality: agribusiness dominates U.S. agriculture.

Part 4

Federal Subsidies to Agribusiness

.

Tax-Loss Farming

We have a government that preaches free enterprise and cultivates monopoly.

—Murray Kempton

In a market economy the federal government and business view social problems from essentially the same perspective. Federal tax laws, farm programs, and agricultural research inevitably support large-scale producers and strengthen the grip of agribusiness. Tax-loss farming—or how to lose at farming and still make a profit—is one source of federal subsidy to corporations and wealthy urban investors. Rather than work the land—a hazardous business in the best of years—they milk the tax laws. In detail, this practice is complicated, but in principle, it is simple: lose money in farming and write those paper losses off against real nonfarm income.

Since the Revenue Act of 1916, the federal government's tax laws have provided greater advantages to nonfarm investors and large-scale producers than to small-scale family farmers. Most of the special tax legislation has been justified as making life easier for farmers and assisting them with the higher risks of raising crops and livestock. Many income tax rules that make tax-shelter farming possible are not specially designed for farming; farming is simply a convenient way of taking advantage of these laws. But, as in other governmental policies, in practice the benefits of these tax loopholes go largely to those least in need of special assistance. High-income doctors, lawyers, politicians, movie stars, and athletes enjoy winning by losing in agriculture. Tax-shelter farming is a rich person's game. Newspaper advertisements soliciting investors for agricultural schemes specify that no one need apply whose tax bracket is less than 50 percent. So while the rich get richer, family farmers are competitively disadvantaged. Agricultural markets are distorted, the public treasury is avoided, land values are artificially inflated, and consumers are faced with higher taxes and higher food prices.

Tax-loss farming is based on two general provisions: postponement of tax payments and reduction of the tax rate. In principle, all kinds of farmers—small and large, part-time and full-time, operating farmers and Wall Street investors—are equally eligible to use tax rules to their own advantage. However, the deductions and concessions have their greatest attraction and impact for high–tax-bracket investors, who have sizable nonfarm income in addition to their farm operations. These investors benefit most because tax rates are graduated and because losses in farming are deductible from nonfarm income. The result is that by reinforcing the unequal distribution of income and wealth in the United States, the tax laws encourage large-scale farming and support agribusiness.

Specific Tax Rules Encouraging Tax-Loss Farming

Only tax rules that relate directly to farming and that are used by persons with farm incomes are considered here. Three major features of the tax code allow wealthy investors to obtain substantial windfalls: cash accounting, deductible capital expenditure, and capital gains. Although investment tax credit, accelerated depreciation, and leveraging are less important, they also provide significant benefits.

Cash Accounting

Whereas all nonfarm businesses must use the accrual accounting method, farmers can choose—and they usually do—the cash accounting method. Under the accrual method, taxpayers must inventory their goods held for sale at the end of the year and add the value of these goods to the total sales from the year, subtracting last year's gross income. Under this method, an expense is effective at the time the *goods* purchased actually change hands. By contrast, farmers using the cash method are not required to keep inventories. Their income is computed on the basis of cash actually received during the year from the sale of products. An expense is incurred the moment the *money* changes hands, even though purchased goods may not be delivered to the farm until the next year.

Cash accounting simplifies bookkeeping records and gives bona fide farmers some flexibility in adjusting their year-to-year income in response to changing weather and market conditions—and what is good for working farmers is even better for tax-loss investors. For tax-shelter investors, who generally are able to afford accountants and bookkeepers and who would normally use the accrual method, cash accounting saves money on bookkeeping expenses and, more importantly, allows premature deductions of expenses against high nonfarm income. This allows the postponement of paying taxes on that percentage of their incomes equivalent to the amount of their farm deductions. These "farmers" effectively receive interest-free loans from the government on behalf of all other taxpayers.

City investors, whose investment cash is essentially surplus, have two advantages over farmers, whose investments provide for their livelihood. First, if profits are realized when crops are sold, tax-shelter "farmers" can reinvest the total amount in another tax-loss venture to earn even more. Farmers, on the other hand, must live off as well as pay taxes on their profits. Second, if actual farm profits are not realized, city investors still gain because

Figure 8.1. Three pillars of governmental power: (top) the president in the White House, (middle) the Congress on Capitol Hill, and (bottom) the justices at the Supreme Court.

of the savings realized from their lower tax bills on their nonfarm income—they can lose and still win! But for working farmers real incomes and farm incomes are identical.

Benefits to wealthy investors are compounded, since the greater the investor's incomes are, the greater the value of each deductible dollar is. The actual subsidy received by tax-loss investors increases in proportion to their income bracket. For example, investors in the 50 percent tax bracket would normally pay half of every $1,000 of income in taxes. If they can deduct a $1,000 feed expense from their tax bill, however, they have in effect paid only $500 for the $1,000 worth of feed, the difference between what they would have given up in taxes and the actual price of the feed. On the other hand, the average farmer's income tax bracket is around 20 percent. Farmers in this tax bracket would save only $200 on a $1,000 feed bill.[1] The richer you are, the more help you get from the income tax code, passed by Congress, to become even richer.

In the 1976 Tax Reform Act, Congress did decide to eliminate the worst aspects of tax-loss farming by preventing corporations and partnerships with corporate partners from taking advantage of the cash accounting method. In order to retain the favorable tax treatment for genuine small farmers, Congress exempted farms with sales of less than $1 million, farms owned more than half by a single family, and farms incorporated as business corporations. Despite the very generous exemptions, two giant chicken producers in Maine and Arkansas complained to their respective senators, Ed Muskie and Dale Bumpers, about the competitive disadvantages they were suffering because of the change in accounting methods. Muskie and Bumpers responded to the cry for help from agribusiness and introduced an amendment to delay for one year the effect of the 1976 changes for these two producers. The debate on the Senate floor shows how persuasive, wasteful of tax revenues, and unfair to small-scale producers are arguments for agribusiness.

> MR. BUMPERS. But the real clinker in it was that the Ways and Means Committee of the House put a provision in which said that any business 50 percent or more of which is owned by one family—and one family is defined in the bill—will be exempt. Here is the net effect of that: you have a business, for example—and let us take Cargill, and I do not know whether Cargill is a family-owned business or not, it is not traded and it could very well be owned by one family—and they do over $5 billion a year, and are big in my State as they are in many States—but if Cargill is, in fact, a family business owned by one family they continue to use the cash accounting system whereas somebody, for example, who started out in the poultry processing business 15 years ago literally selling chickens in a wheelbarrow and they get up to a fairly respectable size of several million dollars a year, they are not exempt and they must go to the accrual accounting system at a very distinct disadvantage.
>
> Now, where is the justice in that? Yet we are saying to some of these other people who are relatively small—and I am talking about the 30th or 40th as poultry processors who are doing somewhere between, say,

$20 million and $50 million a year, just a fraction of what some of these people who are going to be exempt are doing, where is the justice in that?

Mr. President, I want this body to understand one thing: Arkansas is indeed the biggest broiler processing State in the Nation. These are all friends of mine. I support and promote the broiler processing business. It is big business in our country. It is a good cash income for many farmers. All I want is for everybody to be treated alike. Either put them on the cash basis or the accrual basis, I do not care which. They would prefer the cash system as they have always been on it, and I prefer the cash system. But let us not discriminate against some and put them at a very distinct competitive disadvantage for no earthly reason.

MR. KENNEDY. What we cannot get away from, Mr. President, no matter how we describe it, no matter how it is interpreted, is that in the Arkansas situation, we are not talking about a small family farm. What we are talking about is a huge chicken broiler firm that is doing $65 million of business a year in sales, processing 42 million chickens. It is the 24th largest firm in the industry. We are talking about a tax benefit which will be worth $1 million to a single company that has sales of $65 million.

Of the 36 top broiler companies in this industry, 8 were on a cash basis and are going to have to switch to accrual.

Mr. President, the amendment of the Senator from Arkansas and the Senator from Maine selects two of these eight firms and says, "We'll defer this reform for you for a period of 1 year." We closed a major tax loophole for farm tax shelters last year, but we left an exception for some firms; the exception was intended to be for small firms. And now two of the largest firms in the country are coming to Congress complaining that they are too big to fit through the loophole any more. They fall outside the exception we set.

They complain that some of their competitors still fit through the loophole and can use cash accounting, because they qualify for one of the exceptions in the 1976 act. But what about the other six firms who are left behind, if we grant the loophole to these two firms; can they not complain, too?

Why do we not simply close the loophole for all of these giant firms?[2]

Both the Senate and House passed the amendment, providing special exemption for two multimillion-dollar agribusiness firms.

Deductible Capital Expenditures

Under the cash accounting method farmers can deduct expenses of materials and services that actually go into or are part of final salable products—for example, feed, seed, stud fees, and farm management services. Other farm inputs such as machinery, equipment, and improvements to barns are classified as capital assets and are not immediately deductible. Costs of maintenance, upkeep, and development of capital assets are referred to as capital expenditures. For nonfarm businesses capital expenditures are not immediately deductible from taxes. However, under the Revenue Acts of 1916 and 1919, farmers can deduct certain capital expenditures such as the costs of raising livestock held for draft, breeding, or dairy purposes and the costs involved in developing vineyards, orchards (except citrus and almonds), perennial vegetables, and flower beds.[3] Furthermore, within certain limits, expenditures for clearing land and for certain categories of soil and water conservation and fertilization can also be deducted immediately.

The tax consequences of this ruling are illustrated in the following example. Suppose $100 is invested in an orchard. As this $100 is a deductible expense, and as the orchard is not producing any income, a "farm tax loss" of $100 results for the year. This loss may be deducted against income earned in some other way. If I am in the 70 percent tax bracket, my deduction of this loss against, say, medical income will reduce the taxes on my medical income by $70. In other words, 70 percent of the dollars invested in the orchard will be refunded to me by the federal government at the end of the year. Effectively, this reduces the investment in the orchard to $30. If I had not had any income from my medical practice, the farm tax loss I created by investing in the orchard would be of no use to me. Full-time farmers without off-farm income find themselves in such a situation. Deductible capital expenditures help affluent investors increase their wealth and give family farmers a competitive disadvantage in survival.

Capital Gains Tax

When farm assets are sold, taxes must be paid on any profit resulting from the transactions. According to the Revenue Act of 1942 and later court decisions, farm assets such as trees, vines, and animals (draft, dairy, and breeding) can be treated as capital gains. This special provision allows income from sales of capital assets, which have been held for a specific minimum number of years, to be taxed at rates equivalent to half a person's regular tax bracket.

The rationale behind this privilege was based on the special nature of agriculture. Many farm products, such as grapes, tree crops, and cattle, require a substantial time before investments can return profits. Yet the capital gains treatment of agriculture works out better for investors than for real farmers. Again, the benefits increase proportionately to the taxpayer's income tax bracket (see Figure 8.2). To take advantage of capital gains, farmers must sell off their source of income, which could occur only once in their life, either when they retire or when they quit farming. Farm investors, on the other hand, with little commitment to any particular parcel of land or commodity, can liquidate their interests, turn a profit, and reinvest elsewhere. Since they depend principally on their nonfarm income, they can frequently reap the benefits of capital gains.

The advantages of capital gains to investors with nonfarm income can also be shown for the orchard example. Suppose I sell my orchard in which I have invested $100. If I make the sale at my cost, $100, and if I can report that sale as long-term capital gain, I need include only $50 in my income in the

3.6% of the benefits—$25 million—went to 52% of the taxpayers, earning less than $10,000.

12.9% of the benefits—$91 million—went to 32% of the taxpayers, earning $10,000-20,000.

33.3% of the benefits—$235 million—went to 15% of the taxpayers, earning $20,000-50,000.

50.2% of the benefits—$354 million—went to 1% of the taxpayers, earning over $50,000.

Figure 8.2. Farm subsidies: capital gains treatment of farm income. Total tax expenditures in 1977 were $705 million. *Source:* "Those Ever-Regressive Tax Expenditures," *People and Taxes* 6, no. 4 (April 1978), p. 5.

year of sale. As I am subject to a 70 percent tax rate, my taxes on the income of $50 will be $35. Before the 1969 Tax Reform Act, the maximum tax would have been $25. Notice that I have my $100 back. And I have saved $70 in taxes from the deductible capital expenditure, giving me a total of $170. That total must be reduced by the $35 in taxes I paid on the sale of the asset, leaving me with $135. Although I actually broke even in my investment, I am ahead by $35 because my taxes have been reduced by that amount. This $35 is just like a payment from the federal government. The revenue losses from these provisions are estimated at more than $800 million annually. The nominal closing of loopholes in the 1969 tax laws recovered only about $20 million—about one-fortieth of the revenue loss attributable to farm tax losses.

The Revenue Act of 1978 lowered capital gains taxes from the previous rate of half the rate of other income to only two-fifths the rate of other income. The lower tax on capital gains now makes land an even better speculative investment, because profits from its sale are taxed at a lower rate. Investors have a greater advantage in bidding land away from working farmers, especially young farm families whose modest incomes give them the least to gain from tax loopholes.

Investment Tax Credit

Under the Revenue Act of 1971, the investment credit was made available for purchases of livestock and various kinds of real property, such as feed bins and farm buildings. The investment credit is not simply a deduction from taxable income, it is a credit subtracted directly from federal income tax payments. In 1978 tax law made permanent the temporary 10 percent investment credit enacted by Congress in 1971. Allowing recovery of 10 percent of the cost of equipment purchases in tax savings, this tax policy fosters the purchase of new and therefore generally bigger machinery. In addition to equipment costs, these credits also apply to confinement buildings, allowing investors with corporate hog factories, for example, to recover nearly half of their investments through first-year tax savings. These savings have helped corporate hog factories triple their share of the pig crop in only 4.5 years.[4]

Another deduction is available to farmers for soil and water conservation and land-clearing costs if their land is held for ten years. These land-improvement deductions are particularly attractive to high-income professionals who own rural land for recreational use. For improving the land, investors receive higher profits, increased resale value, *and* tax deductions.

Accelerated Depreciation

Investors in ranching take advantage of the accelerated depreciation rule on certain assets, including cattle and real property. The value of cattle bought to build up or improve a herd can be depreciated rapidly. In fact, the deduction can be much larger than the time depreciation and includes a 20 percent first-year bonus depreciation.[5]

Certain fruits and trees also qualify for accelerated depreciation. The 1969 Tax Reform Act required that the cost of raising citrus trees be capitalized during the first four years of the life of citrus trees. A similar provision was extended to almonds in 1970. Florida Representatives Haley and Herlong tried to get the tax code changed to have the cost of citrus trees also capitalized. They finally succeeded in getting their proposal included in the general tax reform of 1969. This helped Minute Maid, a subsidiary of Coca-Cola, plant approximately 200,000 acres of new citrus trees in Florida.

Becoming Tax-Loss Farmers

High-income individuals invest in farming in three ways. (1) Investors enter into limited partnerships that allow profits and losses to pass directly to individual partners. The partnership itself is not taxed, only the proportion of each shareholder. Feedlots and food distributors, such as Montana Beef Industries and Cal-Maine Foods, often set up partnerships to attract clients with capital for their operations. Railroads, oil companies, and utilities figure prominently in the organization of limited partnerships because they have land available for tax schemes. (2) Investors become legal owners of a plot of land or a herd of cattle. An agency invests the investors' capital and manages the farm operations, and the agency is paid a flat fee per head of livestock or per acre managed or a percentage of the gross sales. (3) Investors make their own investment and management decisions. Farm properties could be operated by paid managers or run by managers on share-lease arrangements, in which owners and managers share profits and losses.

Tax laws allow investors to deduct not only expenses incurred by investors themselves but also expenses (e.g., interest) incurred in borrowing money. For example, if the investor's actual cash contribution is $5,000, and that money is used as collateral to borrow an additional $10,000, the investor may be able to take tax deductions worth two to three times the real cost of her or his investment. This common farm investment technique is called leveraging.

Investment farmers need not ever see their farms or even own more than a percentage of a beef herd. The Internal Revenue Service defines taxpayers as farmers if they own, in part or in whole, any farm assets that are developed for profit. Hence, only corporations themselves and individuals in the higher tax brackets, not individual stockholders, are eligible for tax credits and deductions. Corporations are attracted to farming because of favorable tax laws, especially those allowing substantial amounts of capital investment to be written off. Thus they recover approximately 50 percent of the capital costs by deducting them from other taxable income. The lowering of corporate tax rates in 1978 helped incorporated large-scale farmers with taxable incomes in the $50,000 to $100,000 range, and corporations with $100,000 net income realized a tax saving of over 20 percent. This tax feature encouraged big production farmers to incorporate and gave incorporated farmers further incentive to expand—that is, to increase their tax savings.[6]

Extent of Tax-Loss Farming

Although no detailed and comprehensive data are available on tax-loss farming, some general trends can be detected. A clue to the prevalence of tax-loss farming is the large number of tax-shelter promotions and services that are aggressively advertised (see Figure 8.3).[7] Case studies provide one way of documenting the importance of tax-shelter farming. A study at Texas A & M University showed that 90 percent of the 1.4 million head of cattle being fed in the Panhandle-Plains region were owned by individuals and groups other than the feedlot operators, which meant a potential investment of around $348 million by tax-loss farmers.[8]

Limited partnerships and corporations in agriculture for purposes of tax shelters increased in the 1970s. In 1969–1971 the Securities Exchange Commission had farm offerings totaling about $280 million of equity in citrus, tree nuts, vineyards, cattle breeding, and cattle feeding. The gross investments were probably much larger because most offerings were not filed with the commission and because all of them appeared to be highly leveraged. Therefore, Charles Davenport concluded that this amount of equity actually represented somewhere between $1.5 and $3 billion in investments. Corporations in farming also take advantage of tax shelters but since less is known about such enterprises, Davenport suggested that corporations received about $50 million of the total $840 million lost to the U.S. Treasury because of special farm accounting rules.[9]

A special tabulation in 1970 by the Internal Revenue Service at the request of the U.S. Department of Agriculture (USDA) reveals the extent of tax-loss farming.[10] In the fairly profitable farming year of 1970, 43 percent of all individuals reporting farm incomes showed a loss. The percentage was up from 34 percent in 1965. Nearly half of the U.S. farmers were losing money, as measured by their IRS returns. More than 90 percent of the 1970 farm-loss returns had a loss of less than $5,000. The large number of farmers

Figure 8.3. Newspaper advertisement for tax shelters in agriculture.

with relatively small losses would be expected, given the large percentage of farmers whose operations are too small to support their families. Yet the report's statement that "in terms of numbers of U.S. taxpayers and amount of nonfarm income reported, the majority of farm loss returns do not appear to be tax shelters" is misleading. If percentages of farm-loss returns by income bracket are considered, then a marked bias appears. A greater percentage of higher-income than of lower-income taxpayers showed farm losses, and these losses were substantially greater than the losses for low-income taxpayers. The 7,512 taxpayers with farming operations whose gross incomes from all sources exceeded $100,000 in 1970 reported a total of $122 million in farm losses or an average of $16,241 per taxpayer.

In the USDA study the largest income bracket was $50,000 or more, yet Angus McDonald divided the $50,000 and over income bracket into five categories.[11] He found that the percentage of taxpayers with farm incomes who took a farm loss is substantial and is greater for each higher bracket (see Table 8.1). His data show that on an individual basis, taxpayers with large incomes are more likely to take advantage of tax-loss farming and to receive greater benefits than medium- and low-income farmers. This bias is not surprising because our tax laws assure that persons with capital, although fewer in number, will benefit more than the vast majority of farmers who depend on their labor for survival.

If farm losses are categorized by the value of gross farm sales rather than by total income (farm and nonfarm), the largest farms have proportionally greater losses (see Table 8.2). So much for the notion that rich people and big operators make efficient farmers. In fact, Treasury Department statistics show that U.S. millionaires get rich through the tax code by being the world's worst farmers. In 1972, 49 millionaires made $4.3 million on farm operations, but 125 other millionaires managed to lose deliberately $16.4 million from farming! In other words, these otherwise very successful businesspeople

TABLE 8.1
Farm Loss by High-Income Taxpayers, 1970

Income Bracket	Taxpayers		
	Involved in Farming	With Farm Loss	Percent with Farm Loss
Over $1 million	109	90	83
$1 million - $500,000	264	208	79
$500,000 - $200,000	1,554	1,154	75
$200,000 - $100,000	5,595	3,380	68
$100,000 - $ 50,000	23,711	14,647	62

Source: "Tax Loss is Our Loss," People and Land, Summer 1974, p. 7.

TABLE 8.2
Distribution of Farm Loss Returns Within Value of Farm Sale Classes, 1970

Value of Gross Farm Sales	Total Returns (1,000s)	Size of Farm Loss in Percent			
		Less than $5,000	$5,000-$9,999	$10,000-$49,999	$50,000 or more
Less than 2,500	808	95.7	3.1	1.2	0
$2,500 - $4,999	165	92.3	5.5	2.1	0.1
$5,000 - $9,999	122	85.9	9.9	4.2	0
$10,000 - $19,999	84	78.3	15.1	6.4	0.2
$20,000 - $39,999	44	69.8	15.4	13.9	0.9
$40,000 - $99,999	24	51.6	19.6	26.3	2.5
$100,000 or more	8	37.6	17.0	35.2	10.2
All returns	1,255	91.0	5.7	3.1	0.2

Source: Special tabulations by the U.S. Department of the Treasury, Internal Revenue Service, from the 1970 Sole Proprietorship Tax Model in Thomas A. Carlin and W. Fred Woods, Tax Loss Farming, USDA, Economic Research Service Report No. 546 (Washington, D.C.: Government Printing Office, 1974), p. 14.

"lost" an average of $132,000 each in running farms. In the paradoxical world of tax shelters, "going broke" for tax-loss farmers really means "making hay."

As shown in Table 8.3, farm losses are found among all types of crop and livestock farms. Based on the total number of farm-loss returns, livestock farming has the greatest number of returns with losses (52 percent) and field crops the second highest (29 percent). Animal specialty (4 percent), fruit, tree nut, and vegetable (5 percent), and miscellaneous types (10 percent) have proportionately fewer returns. These figures support the claim that the majority of farm-loss returns are submitted by working farmers. Interestingly, these same data also show that the types of farming with small numbers of loss returns have much larger percentages of farm losses as well as higher absolute losses. The rules of the revenue code and court decisions make these latter types of farming attractive for tax-shelter investments. Orchards, vineyards, cattle breeding, and cattle feeding are most attractive because several tax loopholes can be employed.

Tax-loss farming is concentrated in California and the adjacent northwestern and southwestern states. California had 20 percent more farm-loss tax returns than the national average of 43 percent. The Southwest and Northwest had 9 percent and 7 percent more, respectively. Whatever tax shelters investors choose, the reasons for investing capital in a specific farm product have little to do with market demand, soil suitability, or production efficiency, which are the major concerns of working farmers. Rather, investors make their decisions on the potential for maximum tax write-offs, a practice that poses serious social problems.

TABLE 8.3
Farm Loss by Type of Farm, 1970

| Type of Farm | Taxpayers (1,000s) | | Percentage Farm Loss by Type of Farm |
	Involved in Farming	With Farm Loss	
Field Crop	1,059	366(29%)	35
Fruit, Tree Nut, and Vegetable	129	66(5%)	51
Livestock	1,511	658(52%)	44
Animal Specialty	54	47(4%)	87
Miscellaneous	153	118(10%)	77
Total	2,906	1,255(100%)	

Source: Thomas A. Carlin and W. Fred Woods, Tax Loss Farming, USDA, Economic Research Service Report No. 546 (Washington, D.C.: Government Printing Office, 1974), pp. 15-16.

Effects of Tax-Loss Farming

The effects of tax-loss farming are unfair competition to bona fide farmers, higher land prices, overproduction of certain crops, expansion to larger-than-efficient farm sizes, absentee ownership, and the demise of family farms. Those who accumulate various tax subsidies do not have to rely solely on the land to produce an acceptable return; rather, they combine the income produced from the land with tax subsidies. Thus, subsidies have the effect of driving out those who rely on income from the land. When tax-loss investors pay more for an acre than its projected yield warrants, the price of land is driven up, making entry into farming for young farmers more expensive and usually impossible. Existing farmers, without nonfarm income, are also financially squeezed by these higher land prices. Tax-loss farmers effectively make the working farmers' costs rise to meet theirs.

Tax laws also encourage the wasteful use of natural resources. In the Sandhills of Nebraska, for example, investors are installing center-pivot irrigation systems stretching across the tops of hills so steep that they resemble canyon walls. As a somber-faced central Nebraska farmer said: "They've got to be using it for a tax write-off." Indeed, nearly 40 percent of the purchase price of irrigated Sandhills land can be recovered in state and federal tax savings by high-income investors. Irrigated Sandhills farmland is especially amenable to investment credits, depreciation on irrigation equipment, and deductible development expenses because more than half of the $800–per-acre value is in the irrigation well and equipment.[12]

Investments in vineyards and tree crops for tax-shelter purposes can result in overproduction. Since these commodities face an inelastic market demand (consumers can consume only so much fresh fruit, for example, regardless of how low the price goes), oversupplies depress prices and have disastrous effects on small producers, who have no outside incomes to tide them over. Because of tax write-off advantages, investors are less concerned about low prices for their commodities.

To the extent that subsidized investments come in big amounts—and the bigger the investments, the bigger the subsidies—tax loopholes encourage large-scale operations. In California the entire grape acreage in 1972 consisted of 400,000 acres, of which 93,000 was planted in the previous three years—with 53,000 acres in 1972 alone. Two partnerships alone were projected to plant 50,000 acres. This spectacular growth is due to tax shelters and to the increased consumption of wine in the United States. Similarly, tax-subsidized investments give large cattle feedlots, which custom-feed under limited partnerships, advantages over farmer feeding operations. In this way, farm tax loopholes "shift control away from traditional farmers, such as farmer cattle feeders."[13]

The capital involved in tax-loss farming is absentee capital. Absentee ownership of agricultural resources adversely affects rural communities. Since managers are concerned with their investors' short-term profitability, they are less likely to be concerned with the social consequences of their decisions than if they managed their investments as full-time working farmers.

In summary, tax-loss farming allows corporations and high-income persons to receive preferential treatment from the Internal Revenue Service, solely because of their large incomes. The rich become richer and working farm families are placed at a disadvantage in making a living. Bona fide farmers must compete with investors who are not even farming for profit. Rational decisions for relatively few farm-loss investors create irrational market conditions for large numbers of family farmers: artificially high land prices, lower commodity prices for producers, and more absentee and large-scale producers.

U.S. farm tax policies do help individual family farmers at any given time, but in the long term these policies help large-scale farmers more and thereby reduce the ability of any one family farm to compete successfully. Such tax benefits help create and reinforce inequities in agriculture. Because selective consequences of farm tax loopholes affect who can survive, they work indirectly to aid nonfarm investors, corporations, and large-scale producers in controlling increasing portions of U.S. agriculture. Taxpayers pay for these subsidies to agribusiness in two ways, in higher taxes and in the higher prices of an agribusiness-dominated agriculture.[14] Consequently, farm families and consumers in general are not only financing but also absorbing the social and environmental costs of agribusiness dominance.

Federal Farm Programs

The masters of the government of the U.S. are the combined capitalists and manufacturers of the U.S.

—Woodrow Wilson

The purpose of federal farm programs is to restrict production by paying farmers to reduce the amount of land that they cultivate. The programs also seek to prop up the market for crops like wheat, corn, rice, and peanuts by guaranteeing a minimum price. Farmers can collect money for taking land out of production, then increase the yield on the acreage they do use, and collect at least the support price from the government on all that they raise.

Federal farm programs were designed to benefit family-sized farms, but larger-than-family farms benefit disproportionately. The absurdity of federal commodity programs is letting the same rules that established the inequality serve as the allocation model for the administration of benefits—namely, farm size and production. The larger the producers are, the higher their total yields are and the more public assistance they have received. Thus, the net impact of public policy from the 1930s to the 1970s was to encourage farm size increase. Whereas farm subsidies have been defended to the nation as a means of preserving the political and social values of small-scale family farming, this kind of farming has actually declined. Even Don Paarlberg, director of agricultural economics at the U.S. Department of Agriculture in 1972, said that farm subsidies have been a "big hoax" over the last 30 years.[1]

In the 1930s there were hungry people and desperate farmers, but the U.S. government did not put the latter to work feeding the former. Instead, the United States adopted a policy of planned governmentally subsidized scarcity, which helped drive family farmers and tenant farmers off the land, made people dependent on federal food handouts, and contributed mightily to the growth of agribusiness. Over the years these farm programs have continued the trend toward large-scale farming by disproportionately benefiting these operators.[2]

The farm program is massive in scale and impact. From 1964 to 1968 subsidy payments almost doubled to nearly $3.5 billion, and by 1972 a record $4.3 billion was spent on farm subsidies. The average payments varied from

$1,220 for small farmers to $6,646 for the largest farmers (see Table 9.1)[3] One inadequacy of the farm programs is that they are tied to productive resources—land and/or production—and not to farm income need. The more control and ownership of productive farm resources a farmer (or investor or corporation) has, the greater the benefits from these programs are. Automatically large-scale producers benefit more than low-income farmers from such farm programs.

After the 1920s agriculture substantially increased in specialization and production. Since then the United States has had a "farm problem," which is usually defined as "excess" agricultural production relative to "low elasticities of demand": more food and fiber is produced than can be sold at a profit by agribusiness. The federal government has intervened in this crisis in four different ways: price controls, direct cash payments, acreage allotments, and the soil bank program.

Price Controls

After World War I prices received by farmers fell while the prices farmers paid for seeds, fertilizers, and machinery rose.[4] Oligopolies were extracting higher prices from farmers for agricultural inputs and oligopsonies were paying lower prices to farmers for their agricultural goods. These centralizing forces created a farm crisis. To prevent a massive collapse of the farm sector, the government reacted with the McNary-Haugen Bill of 1927. The bill introduced the concept of support prices to maintain farm income levels. The government established minimum commodity prices, and all production not sold in domestic markets would be bought by the government and sold abroad.

This 1927 bill established the precedence of support prices and remains the basis for government intervention in agricultural marketing. Today, the government sets a price for each commodity and lends that amount to farmers who put their commodities up as collateral. When the market price of the product goes above the government set price, farmers are free to sell at the higher price and to pay off their loans. But when the market price is below the set price, farmers keep their loan money and the government takes the commodities and tries to sell them abroad, use them for lunch programs in the United States, and distribute them through the Food for Peace Program.

The Sugar Act of 1934 and its subsequent revisions in 1937 and 1948 illustrate how the price-support program has been used. The act was designed to aid sugarcane producers in southern Louisiana, to insure an "adequate" flow of sugar in domestic and foreign commerce, and to prevent "disorderly" marketing.[5] To qualify for support prices and subsidy payments, farmers were to pay field workers "fair and reasonable" wages, eliminate child labor, preserve soil fertility, and market no more sugar than the quotas allowed. Sugarcane growers—many of whom operate large plantations originally worked by slave labor—have prospered under the act. While they collected over $90 million a year in subsidies during the 1970s, growers failed to fulfill all of the requirements of the act. The present annual wages for cane

TABLE 9.1
Government Payments by Farm Size, 1974

Farm Characteristics	Large-Scale Farms		Medium-Scale Farms	Small-Scale Farms	
	Largest: Sales of $200,000+	Large: Sales of $100,000- 199,999	Sales of $40,000- 99,999	Small: Sales of $10,000- 39,999	Part-time & Subsistence: Sales of Less Than $10,000
Farms (%)	2.1	4.5	14.0	27.3	52.2
Farm Sales (%)	36.8	17.0	24.7	16.9	4.7
Gov't Payments in Each Size (%)	13.0	12.7	12.6	10.9	8.4
Total Payments ($1,000)	44,272	31,555	68,245	84,037	37,788
Total Payments (%)	16.6	11.9	25.7	31.6	14.2
Average Payments ($)	6,646	2,459	1,677	1,220	765
Total Income from Gov't Payments (%)	4.9	4.6	5.7	7.0	6.3

Source: GAO's analysis of 1974 Agriculture Census from Ed Schaefer, Changing Character and Structure of American Agriculture: An Overview (Washington, D.C.: General Accounting Office, 1978), p. 89, Table 25.

workers are almost $2,000 below the federally established poverty level. Most of these workers live in decrepit housing provided by the growers. The physical layout of today's sugarcane plantations resembles that of slave plantations: a large Greek Revival mansion for the planter family with a row of shacks for the workers and their families near the machine sheds and processing plants (see Figure 9.1). Although wage labor has replaced slave labor, exploitation of "field hands" continues with direct government sanction. The average cane workers have only two years of schooling and their children have diets so poor that at the age of twelve they have bodies of fifty-year-olds.[6]

In 1974 sugar shortages resulted in record high prices. Sugar profits in some cases were up 300 percent, yet 33,000 growers continued to receive subsidies. Price supports are used for many commodities in addition to sugar. In 1970 the government spent $4.65 billion in price-support payments, most of which went to farmers who did not need them to prosper.[7]

Direct Cash Payments

During President Harry Truman's administration, direct cash payments were used to control production. Farmers were paid directly to reduce the acreage planted in certain crops. Wool was controlled this way in 1954 and wheat in 1962. The payments had the unintended and undesired effect of stimulating farm production on fewer acres, causing total production to increase beyond the level that the market could absorb profitably. Furthermore, Tweeten concluded from his research that these subsidies very often redistributed income from relatively low-income taxpayers to well-to-do farmers.[8]

Acreage Allotments

Under the acreage allotment program, farmers are restricted to a fixed acreage that can be planted in a particular crop each year. Their acreage limitation is based on the acreage planted during a previous base period. Farmers vote in their counties to accept either (1) acreage controls with enforced penalties for exceeding them and with high support prices or (2) open production with freedom to grow all they want but with drastically lower support prices. Although farmers espouse the virtues of capitalism, they know that if left to the market they would probably be forced out of farming. Hence, most welcome farm subsidies by voting for farm programs. As farm owners, they prefer the government out of their lives, but as farm laborers, they choose government programs that protect them from the very market forces they support as owners.

Allotment restrictions have been circumvented, particularly by cotton growers, by intensifying production on the allowable acreage. To increase their incomes, cotton farmers abandoned the standard forty-two–inch spacing between cotton rows and planted instead at thirty-six– or thirty-two–inch spacings, thereby increasing the plant population by as much as 20 percent.[9] Skip-row planting is another method that increases yields. Every second or

Figure 9.1. Plantation agriculture. (Top) The power and wealth of this Louisiana plantation owner are reflected in the wrought-iron gate and the Greek Revival architecture of
the mansion. In 1969 this 5,000-acre plantation also included a row of houses (similar to
bottom photo) for fifteen black families, machine sheds, and garages where the large
cotton-picking equipment was kept and repaired. (Bottom) "Slave quarters" for black
farm workers near Destrehan, Louisiana, 1938. Russell Lee, FSA photo.

third row is skipped, so that every row receives maximum sunlight and water. In addition, marginal fields with poor soils or steep slopes were taken out of production and the alloted crop grown only on the best land to increase production. The once-extensive Cotton Belt has thus been concentrated in a few already prosperous areas: the Mississippi Delta, the irrigated High Plains of Texas, the Rio Grande Valley, and the Central Valley of California. As a result the intent of the allotment program—decreased cotton production through lowered acreage—was effectively reversed by higher yields per acre.[10] And the benefits from the cotton program were concentrated in the hands of growers with larger cotton allotments and more productive land. The top 10 percent of farmers receiving payments got 48 percent of the benefits in 1961 and 53 percent in 1964. The top 1 percent of farmers in the program received 11 percent of all cotton program benefits in 1961 and 21 percent in 1964—almost double the amount in four years.[11]

In addition, the allotment program has displaced field labor in the old Cotton Belt. Many southern landlords found that the most profitable way to reduce their cotton acreage by 30 to 40 percent to conform with the allotment program was to reduce the number of tenants and sharecroppers by the same percentage. The more prosperous growers could pour on the fertilizer and irrigation water to produce even larger cotton lint yields. The higher yields made mechanical cotton picking profitable for the predominantly white growers but displaced mostly black hand pickers. Federally subsidized research at land-grant colleges on the mechanical cotton picker became profitable for affluent growers when the federally subsidized cotton allotment program took affect.

Soil Bank Program

> *The Bible says a man should work.*
>> —Grower who received $18,000 in federal agricultural subsidies for not planting crops, when asked why he is opposed to federal assistance to the poor.[12]

The soil bank program had two purposes. (1) Farmers "rented" land to the government and in return for not growing crops on this land were paid a subsidy. Under the 1958 National Soil Bank Program, for example, eighteen dollars per acre were paid for wheat land taken out of production.[13] (2) The soil bank program also promoted taking marginal land out of production. During the 1960s absentee farmers in eastern Colorado and western Kansas abused this program by buying cheap high-risk land with no intention of planting wheat but rather to place it in the conservation program and to collect payments for protecting it.[14]

Large and corporation farmers reaped substantial benefits from the soil bank program. A well-documented example is Mississippi's former Senator James O. Eastland. He operated a 5,400-acre plantation in Sunflower County in the Mississippi Delta. Each year the Eastland Plantation produced $500,000 or more of cotton and other crops.[15] Eastland opposed federal government payments to Mississippi's poor, including the fifty to sixty black families that lived on his plantation, but he accepted huge sums of federal handouts for not growing cotton. In 1970, Eastland's cotton subsidy payments were $163,000.

To restrict the abusive gains from the soil bank program, Congress limited federal farm subsidies in 1970 to $55,000 per crop for each recipient per year. This did not stop Eastland and other large growers from receiving more than $55,000. The senator dissolved Eastland Plantation, Inc., and replaced it with a partnership of six entities in which he, his wife, and his four children were partners, and each qualified for the maximum $55,000 payment.[16] Similar legal manipulations allowed other large farmers all over the United States to maintain or even increase their level of subsidy income. For example, in Kern County, California, about one-half of the farmers juggled ownership titles to avoid the $55,000 limitation on subsidy payments in 1971. In the manner of nineteenth-century land swindlers, these farmers also set aside desert land and in many cases bought or leased desert land in order to receive federal farm subsidies. The soil bank program was intended to take out of production high quality land, not to pay farmers for keeping out of production poor-quality land that would have remained idle regardless of the program.[17]

The abuse of the $55,000 limit promoted the following exchange between Senator Taft of Ohio and Congressman Waldie of California:

SENATOR TAFT. As you know, I proposed a $20,000 limitation, not the $55,000 that finally passed.

CONGRESSMAN WALDIE. I do know that. We didn't finally get it in, you are aware of that; we got the words "fifty-five thousand" in, but I looked at the subsidy payments paid this year and I found no one who received less than they got the year before we put the $55,000 in.

That is a fair example of our fantastic inability to do anything against these massively concentrated economic interests, and we tried, and I know how hard you tried when you were in the House of Representatives, but we didn't succeed. We were given a bone thrown to the public because the words "fifty-five thousand dollars" were put in there. There was no $55,000 limit on subsidies; the people in California who were receiving over a million dollars in subsidy received it this year with the $55,000 limitation. Senator Eastland, I am sure, received $150,000 or $250,000 for his cotton farm in Mississippi, as he received prior to the $55,000 limitation.

So I guess what I am really saying is that we do something in the Congress, because we don't have support in the administration, and I don't put it all on your guy. Our administration—

SENATOR TAFT. I tried for many years under the prior administration.

CONGRESSMAN WALDIE. That is what I am suggesting to you, whether he be a Republican or a Democrat, he is unable to move in the interest of this issue. And I don't know the answer to this, except the people have to become aware that their representatives, whether they be in the executive department or the legislative branch, aren't responding to what is in their best interest.[18]

Who Benefits from Farm Subsidies?

James Bonnen, an agricultural economist at a land-grant college, provided a comprehensive analysis of benefits under the major farm programs.[19] Although his calculations were based on data from the mid-1960s, his detailed conclusions on the distribution of subsidies are essentially true today. His data show a significant concentration of benefits from price-support programs among large farmers (see Figure 9.2). With two exceptions—sugar beets and feed grain diversion payments—both price-support and direct-payment benefits from the farm commodity programs were more highly concentrated among large farmers than was farm income. In almost every case, the top 20 percent of farmers receiving payments got more than half the benefits. Conversely, the smallest 40 percent of farmers received only a very modest

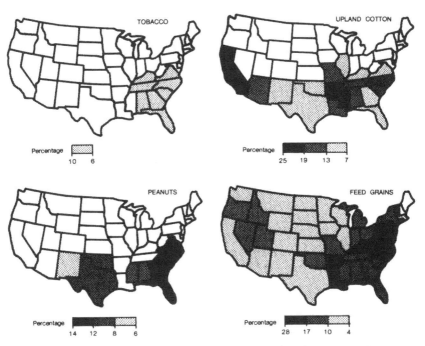

Figure 9.2. Percentage of total benefits received by the top 1 percent of farmers for various crop support programs, 1964–1965. *Source:* James T. Bonnen, "The Distribution of Benefits from Selected U.S. Farm Programs," in *Rural Poverty in the United States*, report by the President's National Advisory Commission on Rural Poverty (Washington, D.C.: Government Printing Office, 1968), pp. 480, 486, 489, and 492.

fraction of the benefits, usually less than 10 percent (see Table 9.2). In 1972, the year direct government payments to farmers peaked, over 41 percent of the subsidies went to the top 10 percent of farmers, and by 1976 the largest 6 percent of farmers received 37 percent of direct payments.[20]

The degree of concentration of benefits among different crop programs, such as sugar, rice, or wheat, depends primarily on the degree of concentration of agricultural production and sales among larger producers. Large-scale producers account for a larger share of sugarcane, cotton, and rice than of wheat, feed grains, peanuts, tobacco, and sugar beets. Economist Charles L. Schultze concluded that "the very nature of current price support programs *guarantees* that benefits will be more heavily concentrated among large farms than is total farm income."[21] This distortion is possible because "price supports raise prices and cash receipts above free market levels by about the same percentage for large and small farmers, but raise net income proportionately more for large farmers than for small ones. And the large farmers' share of total price support benefits will be proportionately larger than their share of net income."[22]

TABLE 9.2
Percentage Distribution of Farm Income and Commodity Program Benefits by Farm Size, Mid-1960s

Source and Year	Lower 20%	Lower 40%	Lower 60%	Top 40%	Top 20%	Top 5%	Gini Ratio[a]
Farmer and Farm Manager Total Money Income, 1963	3.2	11.7	26.4	73.6	50.5	20.8	0.468
Program Benefits							
Sugar Cane, 1965	1.0	2.9	6.3	93.7	83.1	63.2	0.799
Cotton, 1964	1.8	6.6	15.1	84.9	69.2	41.2	0.653
Rice, 1963	1.0	5.5	15.1	84.9	65.3	34.6	0.632
Wheat, 1964							
Price Supports	3.4	8.3	20.7	79.3	62.3	30.5	0.566
Direct Payments	6.9	14.2	26.4	73.6	57.3	27.9	0.480
Total	3.3	8.1	20.4	79.6	62.4	30.5	0.569
Feed Grains, 1964							
Price Supports	0.5	3.2	15.3	84.7	57.3	24.4	0.588
Direct Payments	4.4	16.1	31.8	68.2	46.8	20.7	0.405
Total	1.0	4.9	17.3	82.7	56.1	23.9	0.565
Peanuts, 1964	3.8	10.9	23.7	76.3	57.2	28.5	0.522
Tobacco, 1965	3.9	13.2	26.5	73.5	52.8	24.9	0.476
Sugar Beets, 1965	5.0	14.3	27.0	73.0	50.5	24.4	0.456
Agricultural Conservation Program, 1964							
All Eligibles	7.9	15.8	34.7	65.3	39.2	n.a.	0.343
Recipients	10.5	22.8	40.3	59.7	36.6	13.8	0.271

n.a. = not available.
[a]The more closely the Gini concentration ratio approaches one, the more unequal is the distribution of the benefits. Equal distribution occurs at zero.

Source: James T. Bonnen, "The Absence of Knowledge of Distributional Impacts: An Obstacle to Effective Public Program Analysis and Decisions," in The Analysis and Evaluation of Public Expenditures: The PPB System, a compendium of papers submitted to the Subcommittee on Economy in Government of the Joint Economic Committee, 91 Congress, 1 Session (1969), Vol. I, Table 7, p. 440.

Schultze pointed out a shortcoming in Bonnen's measurements. Although a concentration of 57 percent of price-support benefits among the top 20 percent of feed grain producers, ranked by size of acreage, is significant, knowing their net income is also important. It is inaccurate to assume that an acreage size class is the same as an economic size class. Not all large farms have large sales and incomes and not all farms with large sales and incomes are large acreage farms.[23]

Schultze calculated the distribution of benefits from farm programs by cash receipts for each economic class in 1969 (see Table 9.3). He found that Class I farms accounted for 7.1 percent of all farms but received 40.3 percent of the benefits from these commodity programs. The average net income of these farm operators, from both farm and nonfarm sources, was $33,000, of which 42 percent, or $14,000 per farm, could be attributed to farm commodity programs. Class I and II farms together represented only 19.1 percent of all farms but received 62.8 percent of total benefits. Farm commodity programs contributed $8,000 of their $21,000 average net income. On the other end of the income scale, Class V and VI farms accounted for 50.8 percent of all farms but received only 9.1 percent of the subsidy benefits. The average net benefit per farm was only $400, about 5 percent of their average net income from farm and nonfarm sources.[24] Thus, not only did most benefits go to the richest farmers; in addition, their share of the subsidies was

TABLE 9.3
Distribution of Farm Program Benefits and Income by Economic Class, 1969

Item	Economic Class							
	I	II	III	IV	V	VI	I&II	V&VI
Aggregate benefits	billions of dollars							
Price Supports	1.90	0.76	0.55	0.22	0.08	0.09	2.66	0.17
Direct Payments	1.08	0.90	0.88	0.43	0.20	0.30	1.98	0.50
Total	2.98	1.66	1.43	0.65	0.28	0.39	4.64	0.67
	percent of total							
Total Number of Farms	7.1	12.0	17.0	13.1	9.6	41.2	19.1	50.8
Total Farm Sales	51.3	21.3	16.0	6.3	2.4	2.7	72.6	5.1
Distribution of Benefits								
Price Supports	52.9	21.0	15.4	6.1	2.2	2.4	73.9	4.6
Direct Payments	28.5	23.7	23.2	11.3	5.3	7.9	53.6	13.2
Total	40.3	22.5	19.4	8.8	3.8	5.3	62.8	9.1
Income and Benefits Per Farm	thousands of dollars							
Farmer's Net Total Income	33.0	13.7	9.6	8.1	7.0	8.1	20.9	7.9
Net Income from Farming	27.5	10.5	6.5	3.6	2.1	1.1	16.8	1.3
Price Supports	9.0	2.1	1.1	0.6	0.3	0.1	4.7	0.1
Direct Payments	5.1	2.5	1.7	1.1	0.7	0.2	3.6	0.3
Total Subsidies	14.1	4.6	2.8	1.7	1.0	0.3	8.3	0.4
Net Income from Farming under Free Market Conditions	13.4	5.9	3.7	1.9	1.1	0.8	8.5	0.9

Source: Charles L. Schultze used several USDA data sources to construct this table in The Distribution of Farm Subsidies: Who Gets the Benefits (Washington, D.C.: Brookings Institution, 1971), p. 30.

higher than their share of the presubsidy income. Farm programs actually increased farm income inequality.

Very large individual payments were concentrated even more than benefits in general. In 1968 a total of 261 producers received $100,000 or more; a year later the number jumped to 353.[25] In 1969, a total of 7,795 farmers received payments from the federal government in excess of $25,000, representing an increase over 1968 of 1,881 farmers in this rather exclusive club (see Figure 9.3). These farmers divided a total of $93.3 million more in 1969 than in the previous year. Only a handful of farmers received the lion's share of the subsidies in 1969 (see Table 9.4). In Kings County, California, for example, three agribusiness firms received a total of more than $7.8 million in subsidy payments.[26]

Cotton producers tended to receive larger subsidies, often over $500,000 each in direct payments, than individual wheat and feed grain farmers, who generally received less than $150,000.[27] More than 80 percent of the recipients of payments over $20,000 received their subsidy from the cotton program; indeed, half of the total subsidy payments were for cotton.[28] Not surprisingly, the largest payments in 1969 for crop subsidies went to areas where production was the highest and farms the largest. This was particularly true of cotton. In Kern and Fresno counties, California, over $40 million in crop subsidies were paid to farmers and corporations. On a dollar-per-square-mile basis, the leading county was Sunflower County, Mississippi, with over $13,000 per square mile. Twelve counties, ten in Mississippi, had subsidy amounts of over $7,500 per square mile. Sunflower County is the home of former Senator Eastland, whose plantation received over $146,000 in 1968. Thirteen other farmers in the state received even larger amounts that year.[29]

Representative Silvio O. Conte (R., Mass.) charged that the subsidies make "fat cats fatter," and that payments go to rich farmers in areas with high numbers of poor people. Farm program payments of $5,000 or more were made in 1968 to 2,164 cotton producers in eleven delta counties of Mississippi for a total of more than $49 million. About 40 percent of the families in those same eleven counties were classified as poor, with incomes of less than $400 per family and welfare payments totaling nearly $13 million in fiscal 1969.[30] In Imperial County, $10.6 million of cotton, sugar, and grain federal payments were divided among 252 farmers, or $42,063 per farmer, while the 17,760 poor people in the county received only $439 per capita in federal welfare payments.[31]

The farm commodity programs are irrational and counterproductive to maintaining family farms for five reasons.

1. *Farm price supports and related programs cost taxpayers* an average of $5 billion per year from 1968 through 1970 in federal budgetary outlays. In 1972, the Joint Economic Committee of Congress pointed out that farm subsidies were inequitable and wasteful, not only because taxpayers had to pay enormous subsidies to wealthy farmers but also because the price supports resulted in an additional $4.5 billion in artificially high food prices. The total bill was $10 billion in one year! The total federal, state, and local cost of public welfare programs for the poor, inlcuding Medicaid, was slightly over

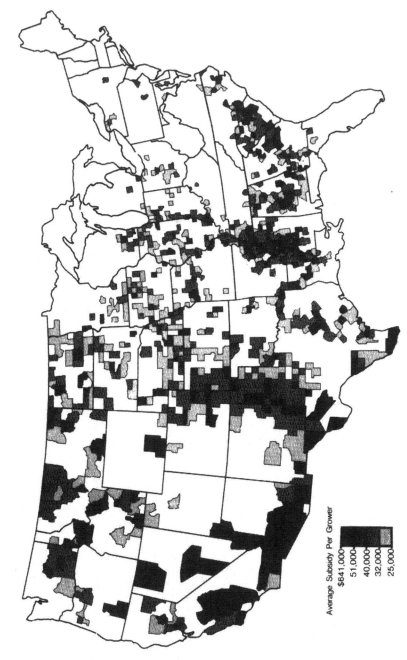

Average Subsidy Per Grower

$641,000
51,000
40,000
32,000
25,000

Figure 9.3. 1969 ASCS program payments of $25,000 or more, excluding price support loans and wool and sugar payments. *Source: Congressional Record*, March 26, 1970.

TABLE 9.4
Farmers Who Received Exceptionally High Farm Subsidies, 1968 and 1969

Location and Beneficiaries	1968	1969
Arizona		
Maricopa County:		
Farmers Inv. Co.	$504,389	$673,410
Pima County:		
BKW Farms, Inc.	331,512	350,607
California		
Fresno County:		
Giffen, Inc	2,772,187	3,333,385
Vista Del Llano Farms	745,647	778,624
Kings County:		
J.G. Boswell Co.	3,010,042	4,370,657
South Lane Farms	1,177,320	1,788,052
Salver Land Co.	786,459	1,637,961
Louisiana		
Natchitoches County:		
J.H. Williams	144,403	158,174
Mississippi		
Bolivar County:		
Delta & Pine Land Co.	605,796	731,772
Leflore County:		
Buckhorn Planting Co.	146,510	180,440
Sharkey County:		
Pantherburn Co.	158,521	194,960
Washington County:		
Potter Bros., Inc.	277,768	248,244
New Mexico		
Curry County:		
John Garrett & Sons	111,799	179,598
Garret Corporation	94,423	147,319
James E. & John Garrett	None	124,899
North Carolina		
Robeson County:		
McNair Farms	351,596	366,584
Southern National Bank	170,044	224,254
Oregon		
Umatilla County:		
Cunningham Sheep Co.	84,993	88,305
South Carolina		
Sumter County:		
W.R. Mayes	156,284	185,089
J.E. Mayes	95,237	122,198
Texas		
Bailey County:		
Carl C. Bamert	96,348	101,763
Bowie County:		
Three Way Land Co.	206,883	198,879
Brazos County:		
Tom J. Moore	289,883	325,886
H.H. Moore & Sons	283,962	310,979
Castro County:		
Hill Farms	174,815	130,979
Jimmy Cluck	142,345	172,359
Cochran County:		
J.K. Griffith	320,315	322,355
Dawson County:		
Bill Weaver	136,105	140,462
Hale County:		
Ercell Givens	156,583	172,152
I.F. Lee	131,881	183,638
Hidalgo County:		
Engelman Farms	155,011	--
Reeves County:		
Worsham Bros.	176,036	222,487
Washington		
Columbia County:		
Boughton Land Co.	140,695	141,688
Whitman County:		
Glenn Miller	116,844	123,532

Source: Paul Findley, "Farm Payments over $25,000," Congressional Record, 91st Cong., 2d sess. (March 26, 1970), p. 9632. When Congressman Findley asked the USDA to tabulate a more recent list, they would not without payment. The earlier list had been made available without cost to Findley.

$10 billion in 1969, no more than federal agricultural welfare for the wealthy.[32]

2. *Most subsidies farmers receive depend on the size of their acreage allotment or their production*, both of which vary directly with farm size. These payments are vested in farmers not as workers or farm operators but as landowners. Farm operators benefit from commodity programs solely on the basis of their control or ownership of land and productive resources. An ever larger share of income, in the form of farm subsidies, goes to the relatively fixed and unequally distributed ownership of farmland. Increasing income inequality is based solely on property rights rather than on the labor contribution to agricultural production.

3. Conversely, *no money from these programs goes to farm workers*, who own nothing but their labor power. In fact, farm programs hurt farm workers by keeping land out of production, thereby reducing the need for labor and increasing unemployment. According to one observer, "The State pays the owners of farm property not to produce, but pays virtually nothing to farm workers who become unemployed as a result of this dole to property owners."[33]

Farm programs, and the soil bank program in particular, have often contributed to increased malnutrition and hunger in rural areas.[34] CBS's documentary film, *Hunger in America,* shows how a poor black family, wanting to raise its own corn to feed chickens, was prevented from feeding itself when their landlord placed the land in the soil bank program. He received payments from the federal government *not* to grow food, and the black family, prevented from using the land to feed itself, was forced to rely on the totally inadequate food stamp program (see Figure 9.4). In short, Congress finds it appropriate to pay farmers for not growing crops but not to pay humans for not working. Herbert J. Gans, an urban sociologist, explained that the American economy "can produce relative affluence for perhaps two-thirds of the population because it is free to exclude or underpay the remainder and pass the burden for their upkeep to the government."[35] In rural areas, the underpaid are mostly working farm families and the excluded are the rural poor (see Figure 9.5).

4. *Government farm subsidies drive up farmland prices.* Because of the income floor provided by the subsidies, land prices are bid up to reflect the land's earning potential. The chief beneficiaries have been those who purchased land before subsidies became prevalent, not young farmers or those expanding their acreage. Farmers who have rented land find many farm-program benefits disappearing in higher rental payments to landlords. In Schultze's words, "Indeed, at the limit, a subsidy attached to land eventually ends up granting no benefits to farmers. . . . [A]fter a number of years have passed, such programs end up transferring little net income to the second generation of recipients, but at the same time become so frozen into asset values that their removal would bring substantial hardship."[36]

5. *High and stable commodity prices stimulate production*, particularly through a faster acceptance of new techniques and a higher rate of investment than would otherwise have been the case. Larger production is counterproductive to the intent of farm programs—which was based on the proposition

Your $10 gift can provide $400 worth of food for the hungry.

On the back roads of Mississippi, Alabama and other parts of the deep South, there are still many thousands of families facing slow starvation. Right now. Right here in the U.S.A.

Their diets are so inadequate that hunger and malnutrition have become part of their lives. Many children of tenant farmers and seasonal workers have actually never known what it is like *not* to be hungry.

The NAACP Emergency Relief Fund is now in its fourth year of collecting money to buy Food Stamps for the neediest of these families. Under the federally sponsored Food Stamp Plan, $1 buys as much as $40 or more in Food Stamps. Thus your $10 can mean $400 worth of urgently needed nourishment to help a family survive.

To contribute to this fund, please send as little or as much as you can to the NAACP Emergency Relief Fund. Contributions are tax-deductible.

Thank you.

NAACP Emergency Relief Fund
Dept. A3, Box 121, Radio City Sta. New York, N.Y. 10019

Figure 9.4. Magazine advertisement in the early 1970s to raise money to purchase food stamps for rural blacks in the South. Reprinted by permission of NAACP Emergency Relief Fund.

that the growth of farm productivity so far outstripped the growth in demand for farm products that price supports and acreage restrictions were required.[37]

In summary, the net results of the farm programs are not surprising: Farm workers are not helped at all, poor farmers are helped slightly, and rich farmers are helped a great deal. Thus farm programs continue a consistent pattern of exploiting family farmers throughout U.S. history.

Class Interests, Congress, and Farm Subsidies

The massive federal outlays in the farm programs, despite the decline in farm numbers, is based on the continued strength of the family farm myth and the farm vote in Congress. The farm vote is strong in urban America for several reasons. First, with each state guaranteed at least two senators, the Senate is permanently gerrymandered to give disproportional representation to farm states relative to their population numbers. Second, the seniority system in Congress has helped preserve farm power. Long-tenured southern Democrats, who are routinely reelected term after term, have secured key committee positions. And third, farm associations—representing conservative farmers or farm trade groups—spend a great deal of money lobbying in Washington. Of the top twenty-five spenders in the capitol in the early 1970s, four were farm groups: American Farm Bureau Federation, National Farmers Union, National Council of Farmer Cooperatives, and National Cotton Council.

Why does Congress pass farm legislation that consistently favors well-to-do farmers and largely harms modest- and low-income farm families? The answer is not hard to find. It takes money to get elected to Congress; the

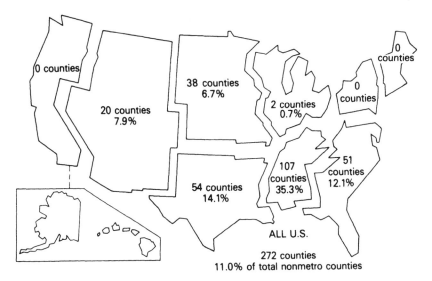

Figure 9.5. Nonmetropolitan counties with less than $3,500 per capita income, 1975. Percentages refer to the proportion of total nonmetropolitan counties in a division.

greater the financial resources one has, the easier it is to be elected (especially since the Supreme Court ruling that the Campaign Reform Acts restrict all donors in the amount they can give, but not the candidates themselves, who may spend an unlimited amount in an attempt to be elected).

People of wealth tend to support government farm policies and programs that reflect their own personal and class interests. Ralph Nader's Citizen Action Group showed that the 1975 Senate had at least twenty-one millionaires but only five senators who shared the material condition of 99 percent of Americans—less than $50,000. Wealth and political ideology are strongly related. Americans for Democratic Action (ADA) determined that senators with the lowest assets had the highest liberal voting record (92 percent of all votes cast), while the millionaires had the lowest liberal voting rate (29 percent) (see Table 9.5). There are many individual exceptions (millionaire senators do include liberals like Ted Kennedy or Phil Hart), but conservatives like James Eastland, Barry Goldwater, and Russell Long are more typical of the majority. As Jim Chapin put it: "Can we reasonably expect an assembly whose average member is worth half a million dollars to shift the economic burdens of our society?"[38]

Congress also thwarts the interests of family farmers and agricultural workers by conducting committee meetings in secret. Major decisions, which often are difficult to reverse later on the House and Senate floors, are made in committee meetings. In 1969, 29 and 35 percent of Senate and House committee meetings, respectively, were closed; by 1972, 59 and 49 percent were closed. In both chambers, agricultural committees had become significantly more secretive than the average congressional committee (see Table 9.6).

The composition of the congressional committees on agriculture constitutes another way farm policies reflect bias by emphasizing certain regional crops and large-scale farmers. Economist Arnold Paulsen has pointed out that

TABLE 9.5
Wealth and Ideology in the U.S. Senate, 1975

Wealth Category	Senators in Group	Average Liberal Rating(%) by ADA, 1975
Under $50,000	5	92
$50,000 - $250,000	30	59
$250,000 - $500,000	18	53
$500,000 - $1 Million	4	53
$1 Million or More	21	29

Source: Ralph Nader study, reported in Jim Chapin, "The Rich Are Different...," Newsletter of the Democratic Left, November 1976, p. 3. The results are for 88 senators for whom questionnaire or other data were available; leaving 22 senators for whom no reliable information was available.

TABLE 9.6
Committee Secrecy in the Congress, 1969 and 1972

	1969				1972			
	Meetings			Percent Closed	Meetings			Percent Closed
	Open	Closed	Total		Open	Closed	Total	
Senate								
Agriculture and Forestry	25	10	35	29	15	22	37	59
All Senate Committees	1,085	427	1,512	28	1,041	613	1,654	37
House								
Agriculture	83	45	128	35	39	38	77	49
All House Committees	1,390	1,024	2,414	42	1,307	1,022	2,329	44
All Congressional Committees[a]	2,559	1,470	4,029	36	2,425	1,648	4,073	40

[a]Includes joint House and Senate committees as well.

Source: "Committee Secrecy: Mirror Impacts of Reform Act," Congressional Quarterly, February 12, 1972, p. 302; and "Committee Secrecy: Still Fact of Life in Congress," Congressional Quarterly, November 11, 1972, p. 2976.

the resistance to subsidy dollar limitations in 1970 ($55,000 for each individual per crop instead of the unlimited amount per individual that existed in 1968) came from politicians from a rim of states from Hawaii and California across to South Carolina who held key positions on congressional agricultural and appropriation committees.[39] Indeed, the Agricultural Act of 1970 was passed by Congress only because members of urban districts in the Northeast, Midwest, and California voted against members from the rural parts of the South, Plains, and Rocky Mountains. According to Paulsen, "these latter regions and their representatives have respect for big operators and for property rights and asset values."[40] These areas are also characterized by the concentration of large-scale producers in general and of large-scale sugar, rice, and cotton producers in particular.

The relationship between committee representation and crop subsidies per district is illustrated by the House Committee on Agriculture. Political geographer Stanley Brunn determined that "of the 34 committee members, 29 had districts that received over $4 million in fiscal year 1968."[41] In some districts the payments were primarily for one crop such as cotton and in other districts payments were for wheat and feed grains. The largest cotton payments went to four districts: District 1 in Arkansas, District 18 in Texas, District 1 in Mississippi, and District 18 in California, which received $30 to $60 million. The greatest support for wheat payments went to District 1 in Kansas ($96 million) and District 2 in North Dakota ($54 million). Feed grain payments from $20 to $26 million each went to District 6 in Iowa, District 1 in Kansas, and District 18 in Texas. A total of over $600 million was received by congressional districts whose representatives were on the important House Agricultural Committee. Of the leading ten districts receiving subsidies in 1968, five had members on the House committee. District 19 in Texas, represented by Congressman Price, and District 1 in Kansas, represented by Congressman Dole, received the highest subsidy payments: over $117 million and over $102 million, respectively.

Such bias is to be expected given the particular regional composition of the House subcommittees on commodities. National policies that affect everyone are being formulated by class and regional self-interest. Six of the seven-member cotton subcommittee came from cotton-producing areas. Similarly, representatives from the major livestock- and grain-producing areas and from the principal dairy and poultry regions predominated on their respective subcommittees. On the other hand, only two of the eleven-member subcommittees on family farms, rural development, and special studies came from districts with high levels of rural poverty. The major conclusion to be reached is that these subcommittees overrepresented producers of certain crops and therefore underrepresented all other farmers, agricultural workers, and rural people in general (see Figure 9.6). In areas with particularly high subsidy payments, acute poverty and even hunger conditions exist, particularly in the Mississippi Delta and the Carolina Piedmont areas. In 1968 Mississippi County, Arkansas, for example, with 29 percent of the population below poverty level, received over $12 million in direct cotton payments but only $625,000 for food stamp coupons.[42] Typically, defenders of subsidies

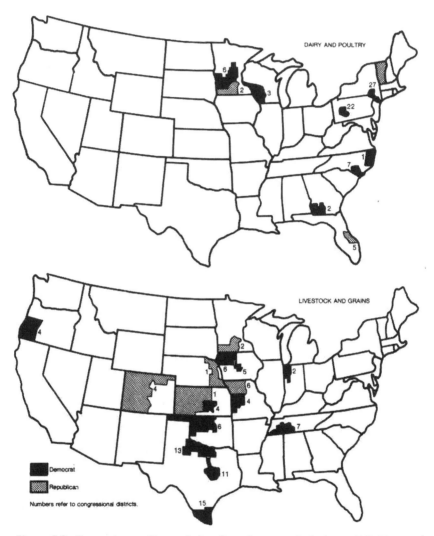

Figure 9.6. Four subcommittees of the Committee on Agriculture, U.S. House of Representatives, 1977.

tell consumers that the main purpose was never to help the rural poor and small-scale farmers but to stabilize production for the benefit of consumers. In practice, as Schultze has shown, the farm subsidies have encouraged intensification of production and higher total yields and increased the cost of food to consumers.

The common argument that committee members support the farm programs solely to get reelected obscures more than it reveals. How can one

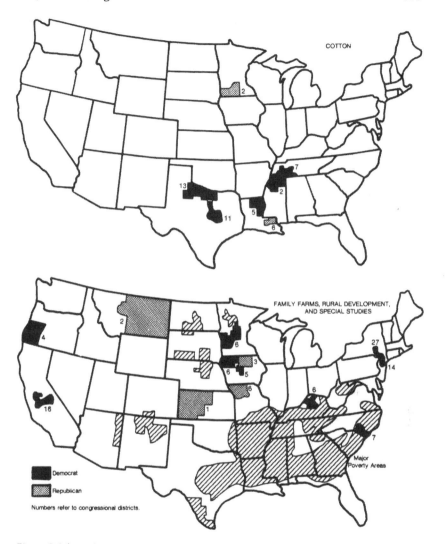

Figure 9.6 *(cont.)*

speak of seeking votes from one's farm constituents when most of the benefits from the farm programs go to a small percentage of farmland owners and none go to the many landless tenants and agricultural workers? By claiming to benefit all farmers equally, such policies receive support from other members of Congress and, more importantly, obscure their hidden distributional bias. Small-scale farmers, in particular, can pretend that farmers are being treated equally and fairly. Ironically, the very farmers who as a result of the farm programs face unfair competition from the heavily subsidized large-scale farmers support the politicians whose legislation furthers the demise of family farms.

The appeal of equal treatment before the law is used to perpetuate and even, in many cases, to exacerbate farm income differences. The class interests of property and capital are camouflaged by the rhetoric of democratic principles. As a result, family farmers, who have little to gain and a great deal to lose from current farm programs, continue to support politicians and a legislative system that is bankrupting more and more of them each year.

Farm Programs in the 1970s

In 1972, Earl Butz, Secretary of Agriculture in the Nixon administration, nearly dismantled the entire federal farm program, which had been virtually unchanged since the 1930s under Franklin Roosevelt. During 1973 he convinced Congress to take away the authority he held under the Agricultural Adjustment Act to encourage farmers not to grow food and fiber. Butz sold off all government-owned storage bins, which held the "surplus" that the government had bought over the years. His 1973 bill, which expired in 1977, removed the government from agricultural marketing entirely. The law compensated farmers only if prices dropped disastrously and only at a fraction of their production costs. By 1974, every acre of arable land was being used, farm income and prices had reached record highs, and grain exports to the Russians were paying for petroleum imports.

Butz spared no praise for his laissez-faire farm approach. In 1975 he said: "During the past five years we have moved out of that straitjacket price support program into a market-oriented system, and farmers have enjoyed price levels beyond their most optimistic dream of the 1960s."[43] Farm prices did increase rapidly, but not because of the dismantling of farm programs. Massive worldwide droughts in the Soviet Union, China, India, and Africa temporarily created a large demand and higher market prices for wheat, beef, and other U.S. agricultural products. The conservative ideology of Butz appealed to most farmers during this period of prosperity, but once the global weather conditions improved farmers sought the shelter of government planning again.

Many liberal legislators and liberal farm organizations questioned Butz's interest in helping farmers. His free-market policies were more geared to agribusiness than to farm families. He had been dean of continuing education at Purdue, Indiana's land-grant university, where he earned $35,000 a year. After repeated questioning at his Senate confirmation hearings, he admitted he owned 2,000 shares of stock in Ralston Purina Company and International Minerals and Chemicals Corporation and 1,000 in Stokely-Van Camp Company. The stock was worth about $109,000. In addition, he earned $26,800 annually as a director of these three corporations. His annual income and corporate associations gave him different class interests from the majority of U.S. farmers who had sales of less than $20,000. His callous attitude toward consumers and the poor—whether they were migrant farm workers or recipients of federal food stamps—was also known. If Butz represented any farm interests, they were agribusiness, not the vast majority of farmers.

By the time President Carter took office, the world food situation had changed considerably. Good weather in India, China, and Russia produced bumper crops. After a short interval, the farm economy was again characterized by food "surpluses" and falling farm incomes. The Food and Agriculture Act of 1977 was designed, consequently, to provide stability once again for farm income in an uncertain market. The government would make "deficiency" payments to farmers when the market prices for commodities fell below congressionally established "target prices."[44] For example, the 1977 act established a minimum dairy support level at 80 percent of parity, with semiannual adjustments for inflation. By 1978 the USDA had already paid farmers not to plant 15 million acres. In 1979 Senator Herman Talmadge of Georgia, chairman of the Senate Committee on Agriculture, thought the government should pay farmers about $75 per acre to keep an additional 31 million acres idle. Senators George McGovern of South Dakota and Robert Dole of Kansas suggested simply raising the farmers' incomes by raising the federal "target price" of corn, wheat, and other items, so that the government would pay farmers the difference between the actual selling price of these crops and the ideal price that farmers and politicians agreed upon. Senator Edmund Muskie of Maine estimated that in addition to direct costs to the U.S. Treasury, this farm bill might increase consumer food costs by as much as $6 billion by 1979. These proposals are consistent with past policies in which taxpayers and consumers pay the costs and the private sector, particularly agribusiness and land-wealthy farmers, reaps the benefits.

Federal Milk Marketing Orders

Milk, with its perishable, bulky, and seasonal qualities posed special problems for dairy farmers in the early twentieth century. Forced to sell their milk quickly and close to home, farmers had little bargaining power over the prices they received. Consequently, in the early 1900s milk producers formed cooperatives to offset the oligopsonistic power of milk dealers and to promote market stability. During the Great Depression chaotic market conditions alarmed the cooperatives and they sought help from the federal government. Thus, the modern-day regulatory framework for milk and dairy products is rooted in the Agricultural Marketing Act of 1937.

Federal regulation of milk markets operates through federal milk marketing orders, which were established in the 1930s. A milk order is a regulation, issued by the secretary of agriculture, setting minimum prices farmers will receive and dairies will pay for raw Grade A milk in specific regions. At last count there were 52 federal orders regulating about 80 percent of Grade A milk, which meets the sanitary requirements for consumption. Orders classify Grade A milk into either Class I or Class II, depending on use. Class I milk is used for fluid consumption; Class II milk is used in manufacture of dairy products such as cheese, yogurt, and dry milk. Some Grade A milk ends up in manufacturing uses because of federal milk pricing regulations, which stimulate total Grade A milk production beyond fluid milk market needs.

The surplus gets channeled into manufactured milk products.

Grade B, used for manufactured dairy products only, is not explicitly regulated. Its price is set either by market forces or by government price supports. The influx of Class II milk in this second market pushes down cheese prices. Support prices then come into play to keep them from falling too far. The federal government's Commodity Credit Cooperation (CCC) purchases manufactured dairy products in order to prop up prices in this market. On September 1, 1978, the CCC held 291 million pounds of butter and 40 million pounds of cheese. The federal government commonly buys 5 percent or more of total U.S. milk production during a year to prop up prices.[45]

Several estimates have been made to determine the size of the benefits to dairy farmers and the costs to consumers. Although the USDA argues that the program results in "modest" enhancement of Class I milk prices, the Federal Trade Commission (FTC) suggests that fluid milk prices are from 4 to 15 percent above competitive levels, depending upon the year.[46] The FTC has also examined the distributional benefits of the federal milk marketing orders. The absence of precise, current net-income data restricts the FTC's study to the distribution of gross program benefits.

The value of farm product sales is a substitute for bigness and wealth. The property and asset characteristics for dairy farms by economic class demonstrate that the largest dairy farms are the wealthiest. Class I dairy farms represented 13 percent of all dairy farms in 1969 but accounted for 25 percent of the land in farms, 34 percent of the value of land and buildings, and 41 percent of the total value of all agricultural products sold. They were also the largest in terms of average acreage and average number of milk cows. On the other hand, Class V dairy farms represented 8 percent of all dairy farms but accounted for only 4 percent of land in farms, 3 percent of the value of land and buildings, and 1.2 percent of the total value of agricultural sales.[47]

Although the milk program provides at least some benefits to all dairy farmers, larger farmers receive a much larger share of the benefits (see Table 9.7). In 1969 Class I farms represented 13 percent of all dairy farms but received 41 percent of all federal milk order benefits. Classes IV and V, on the other hand, represented 25 percent but received only 3.5 percent of the total benefits. Class I farmers, the wealthiest, received $15 for every $1 received by Class V farmers, the neediest. The modest subsidy of $2,000 may be very significant as a proportion of income for the smallest farmers, but they are being helped at the enormous cost to the taxpayer of much larger payments to the wealthy.[48]

The benefits are also unequal regionally. Regions with large dairy farms tend to have the greatest concentration of milk benefits. The South and the West Central regions reflect this trend, while the West and the Midwest have lower concentrations of benefits. The Northeast has the least concentration of benefits owing to the region's smaller-sized farms. The results of the milk programs have provided, according to the FTC report, "a deluge of benefits for large dairy farmers and only a trickle of additional gross income for small dairy farmers."[49]

TABLE 9.7
Distribution of Milk Subsidies by Size of Commercial Dairy Farms, 1969

	Total	Class I Total	Class I Class Ia	Class Ib	Class II	Class III	Class IV	Class V	Gini Ratio
U.S. Total[a]									
No. of farms[b]	259,850	34,153	4,808	29,345	81,668	78,805	45,339	19,885	
Percentage of farms	100	13.1	1.9	11.3	31.4	30.3	17.4	7.7	
Million Dollar benefits	145.4	58.9	15.3	43.6	56.6	24.9	4.9	.1	
Percentage of Benefits	100	40.6	10.5	30.1	38.9	17.1	3.4	.1	.463

[a]Excludes Alaska and Hawaii; and excludes Arizona, Montana, New Mexico, and Wyoming because no data were available.
[b]Also includes those dairy farms which sold all of their milk outside of federal orders. An attempt to extract such farms would have generated unreliable results.

Source: David R. Fronk, Farm Size and Regional Distribution of the Benefit under Federal Milk Market Regulation (Washington, D.C.: Federal Trade Commission, Bureau of Economics, 1978), pp. 29-33.

How the Milk Lobby Works

Like most people in business, dairy cooperatives and dairy farmers may speak highly of free enterprise but have no great desire to experience it themselves. To protect themselves from its "destructive" force, they lobby the federal government for protection. The milk industry is characterized by massive government subsidies, monopolistic practices, and unnecessarily high consumer prices that benefit agribusiness rather than family farmers. The power of the milk lobby was illustrated in 1971 when Secretary of Agriculture Clifford Hardin was persuaded to change his mind about the level of government milk support prices. On March 12, 1971, Hardin announced that the milk support price would be held at $4.66 per hundredweight. Thirteen days later, Hardin announced that the government-guaranteed support price would be raised to $4.93, an increase of 27 cents. Given that administration economists estimated that this increase would cost taxpayers up to $300 million in increased subsidies, how did Hardin's change of mind come about?

The major events between March 12 and March 25 were

1. In Congress 127 representatives and 29 senators sponsored various bills to raise milk prices. Of the 71 sponsors of the increased milk support-price bill, more than 30 received campaign funds from the milk lobby during 1969 and 1970 in amounts ranging from a few hundred dollars to several thousand. For example, Humphrey received $10,523, Proxmire $8,160, Muskie $7,132, and McGee $2,000.[50]
2. On March 22, the Trust for Agricultural Political Education (TAPE), the political arm of the largest dairy co-op, American Milk Producers, gave its first reported money, totaling $10,000, to Nixon's 1972 campaign.
3. On March 23, Nixon met with a dozen executives from the dairy industry and later met privately with administrative officials, including Treasury Secretary John Connally.
4. On March 24, the Trust for Special Political Agricultural Community Education (SPACE), the political arm of the third-largest co-op, Dairymen, gave $25,000 to Nixon's campaign.
5. On March 25, Hardin announced the milk price increase.

Dairy political funds were distributed to all who supported increased milk supports. TAPE, SPACE, and ADEPT (Agricultural and Dairy Educational Political Trust, the political arm of the second-largest co-op, Mid-American Dairymen) eventually contributed $427,000 to Nixon's reelection campaign, $532,050 to Republican candidates, and $616,892 to Democratic candidates.[51] The milk lobby's money was well spent. The price increase was worth an estimated $300 million to the nation's dairy farmers; $75 million of that went to the members of the three big co-ops that donated the political money. These three dairy co-ops, covering an area from the Alleghenies to the Rockies, controlled about 25 percent of the nation's milk. Farmers voluntarily contributed $1.3 million a year to the lobbying effort through check-offs from their milk checks. TAPE spent $906,245 in 1972. That figure made it the second largest political pressure group in America, topped only by COPE, the political trust of the AFL-CIO. But the AFL-CIO represented 13 million people, and TAPE's financial influence came from only 40,000 farmers.

The House Judiciary Committee investigated charges that political contributions by the dairy lobby constituted bribery, but it found no wrongdoing. Of course, sixteen of the committee's members received donations from the same dairy groups, so they were not exactly disinterested.[52] Plenty of politicians continued to take milk money. From September 1973 to August 1974, the dairy co-ops had given $57,473 to Democratic Party committees and $35,540 to GOP committees. The money got passed along to individual candidates, but with the stigma laundered out. Individual politicians also continued to receive direct donations. Senator Herman Talmadge as chairman of the Agriculture Committee received $10,000. On the House Agriculture Committee, $15,000 went to Representative R. Brown (D., Miss.), $5,090 went to Jerry Litton (D., Mo.), and $5,000 to Frank A. Stubblefield (D., Ky.).[53]

Milk price supports increased three times during President Carter's administration. Wisconsin, a dairy state, was a crucial state in Carter's 1976 election. Although Carter's own Council on Wage and Price Stability recommended that the milk support price be lowered, Secretary of Agriculture Bob Bergland announced an unexpectedly high support level representing 83 percent of parity. The new support level in effect redeemed Carter's pledge to dairy farmers and benefited them in the short term. This increase doubled the support program from $300 million in 1976 to $600 million in 1977. In addition, the government also bought $483 million of "surplus" milk powder. Ironically, the dairy co-ops that wanted price-support increases in the past opposed this latest increase. As lobbyist Pat Healy of the National Milk Producers Federation said: "We fear that the shock may make the Congress rethink the funding of the price support program."[54] Through either direct campaign contributions or votes, the milk lobby got the support price increased from $4.66 in 1971 to $9.00 in 1977. This was a 93 percent increase during a period when the general cost of living rose 49 percent![55] By March 1979 the support price had risen to $10.44 per hundredweight.

The consequences of the federal milk program on consumers and family farmers are multi-faceted, far-reaching, and disastrous. The federal milk orders have increased the cost of milk to consumers. Higher prices, along with massive advertising of lower-priced soft drinks—especially diet drinks—resulted in a decrease in per capita annual consumption of milk and cream from 302 pounds in 1965 to 244 pounds in 1974. The current program also results in overproduction of manufactured dairy products, particularly cheese, which is increasingly used in the fast-food industry. The government is thus subsidizing fast-food consumption while discouraging healthier milk drinking.

Because farmers received the highest prices for Grade A milk, they were pushed toward fluid milk production. The increased production costs invariably pushed them toward bigness. And because the program encouraged overproduction of milk, it mandated increasing size. Dairy farms decreased in number from four million to under 200,000 in the last twenty years. But production has increased over the years (even as consumption went down), evidence that the fewer farms were getting bigger and bigger. The general forces toward agribusiness have been felt by dairy farmers, who have been given additional reasons to get bigger by the current milk programs. Since the benefits are distributed disproportionately to large producers, bigness is given yet another helpful hand by the state.

The milk marketing order system has also encouraged the dairy co-ops to become local and regional monopolies, to the detriment of small-scale producers and consumers alike. The growth of co-op monopolies flows logically from the milk market orders because set prices inhibit movement of milk from one marketing area to another in response to demand and competitive prices. The Copper-Volstead Act of 1922 exempts co-ops from this anti-monopoly legislation. According to the Justice Department, "Some cooperatives, making generous use of their Copper-Volstead exemption and the federal order system, have achieved and exercised monopoly power. Without the federal system, none of this would likely have occurred and certainly would not have been allowed to persist."[56]

At the same time that the federal milk program encourages large-scale growth, it also allows the survival of small farms that would go out of business under free market conditions. As Senator Gaylord Nelson of Wisconsin said, "Those family dairy farms that have survived—and many have not— have done so because the present price support program works."[57] The number of family dairy farms in the Northeast and Midwest would be substantially reduced if the milk program were discontinued. The illusion of viable family farming is maintained in the dairy regions, while its demise, should milk programs be terminated, is assured because benefits have helped consolidate the competitive advantage that would be held by the few top producers in each state.

Racial Discrimination in Federal Farm Programs

Most federal farm programs are administered at the county level with farmer representation. Although these programs supposedly help all eligible farmers, regardless of race, the discriminatory practices within the agencies and programs of the USDA have limited minority farmers' participation in these badly needed and publicly supported services and have accelerated the rapid exodus of blacks out of farming. Racism appears in the staffing of USDA programs, participation rates in local agricultural decision-making groups, and utilization of federal farm programs.

Staffing USDA Programs

Minorities make up only 10.9 percent of all permanent, full-time USDA employees, but 20 percent of all federal employees. In Mississippi during 1975, blacks represented 31.5 percent of the rural population, but only 5.1 percent of the state's USDA employees. In the same year, North Carolina had only 7.2 percent black USDA employees, yet blacks accounted for 19.8 percent of its rural population. Minorities constituted only 3.2 percent of the

Figure 9.7. Industrial dairy farming. (Top) Large-scale dairy farm with 240 milking cows and an additional 1,000 cows and calves. This farm in Porterville, California, consists of only 240 acres because hay and silage are brought from as far away as the Imperial Valley. (Bottom) Inside a large-scale dairy barn in Porterville, California. Cows are milked much as in a car wash. One such farm is one hundred times the size of an average Wisconsin dairy farm.

Figure 9.8. Family dairy farming. (Top) Aerial view of a small-scale dairy farm in Stearns County, Minnesota. (Bottom) Interior of a dairy barn with fifteen milking cows. One- and two-person dairy farms still typify the Midwestern dairy region.

8,600 employees of the Agricultural Stabilization and Conservation Service and only 7.2 percent of the more than 17,500 Cooperative Extension Service employees. Legal action was taken as late as 1974 in Mississippi to get the first black appointed as a full-time county agent.[58]

Participation in Decision-Making Groups

Underrepresentation of minorities on the staff and decision-making bodies of the USDA farm programs has systematically prevented black farm families from receiving the same assistance that equivalent-size white-owned farms have received.

Large landowners (all white) in the South dominate the administration of federal agricultural programs. Their political and economic powers extend from Congress to the local administrative offices in the county seats. They control the Agricultural Stabilization and Conservation Service (ASCS), which administers commodity support prices and establishes acreage quotas for basic crops, of which cotton and tobacco are the two most important, especially for black farmers. As recently as 1968 no black farmer in the South had ever served as a regular member of a county ASCS committee.[59] Only eleven of the 9,097 ASCS county committee members in the nation in 1977 were black, and only three blacks were included among the possible 3,857 county committee members in the South. Of the 15,540 supervisors on the district boards for the Soil Conservation Service, only 2.1 percent were minorities. In 1975, the FmHA's county committees included only 11.2 percent minorities.[60]

Utilization of USDA Programs

The Farmer's Home Administration (FmHA) was established in the 1930s to make loans available to small-scale farm families to buy machinery to increase productivity and thus expand their farm incomes. According to one study, "During the period 1964 to 1967, black farmers, who constituted about a third of all farmers in the South, received only a fourth of all loans and only a seventh of the total funds."[61]

Similar discrimination is apparent in the utilization of USDA programs. In the decade ending in 1976, the percentage of blacks receiving housing loans from the FmHA declined from 19.4 to 9.5 percent. Ineligibility accounts for much of the decline; by 1974, 50.3 percent of rural black families were too poor to qualify for the loans (up from 36.2 percent in 1969). White households too poor to qualify, however, remained about the same over the decade—25 percent. Between 1966 and 1976, the percentage of FmHA farm ownership loans made to black farmers declined from 5.7 to 1.5 percent. The ASCS likewise experienced lower minority utilization. In 1975, only 40 percent of eligible black-operated farms participated in USDA soil and water conservation programs while 59 percent of white-operated farms were enrolled.[62]

Conclusion

Contrary to the Supreme Court's reasoning regarding racial discrimination, the intent of a law or practice is not sufficient to judge its merit. Rather, the heart of the matter rests on the law's impact on people. If the intent of Congress was not to discriminate against family farmers, regardless of race, in the various farm programs, this intent does not help the family farmers who have gone bankrupt because agribusiness has been subsidized at their expense. If the outcome of economic and legislative action is unequal, these practices and actions are indefensible in a democracy based on equality and justice. At best, Congress has treated farmers equally, but since farmers have had unequal control and access to resources—land, capital, technology, and information—the outcome has increased that inequality. Thus, under the cover of the myths of equality and of the family farm, U.S. farmers and farm workers have actually been treated unequally—and the resulting rural social problems have been blamed on them!

The Land-Grant College System

In conjunction with the land-granting and irrigation laws of the late nineteenth and early twentieth centuries, Congress passed several laws designed to give farmers their own colleges and a department of agriculture in the federal government. Conservative business interests opposed such "radical" demands, but they also needed to pacify rural areas. The Homestead Act, the Morrill Act (which established agricultural and mechanical colleges), and the establishment of the U.S. Department of Agriculture (USDA) could pacify the demands of farmers as well as benefit industrialists. "To the rising individual leaders of the country, and their political groupings, large expenditures for farm education [Morrill Act of 1862] and research [Hatch Act of 1887] were far better than radical Greenback money policies, inflation, farm strikes, and attacks on the railroads and the industrial trusts."[1]

The Initial Acts

Although the stated intentions of Congress for promoting rural education, agricultural research, and farm extension work were in the best interests of farm people, eastern industrialists, who wanted cheap labor and cheap food to increase their profits, became the major benefactors of increased farm mechanization and productivity. The land-grant colleges, the USDA, agricultural research stations, and the extension service, instead of raising farmers to the levels of professional and business classes, encouraged the heavy injection of new technology into agriculture and new capital to support this technology. These institutions facilitated the growth of agribusiness and thus created inequality where much more equality had existed before.[2]

Morrill Act of 1862

The Morrill Act granted each state 30,000 acres of public land for each senator and representative in Congress. The money from the sale of these lands was to be used to establish agricultural and mechanical colleges. Eastern states lacking the necessary public lands within their borders were issued land scrips in equivalent amounts to be applied in states with public lands. Cornell University in New York, for example, was established on land scrips applied in western Wisconsin, which has a town by the same name.

The first Morrill Act was the initial step in democratizing higher education in the United States. Before 1862, higher education was the exclusive domain of the elite, the rich, and the professional classes. But the government's commitment to the people's colleges was limited and characterized by struggle. Although 234 million acres were made available to homesteaders and 181 million acres were granted to railroads, only 17 million acres were set aside for the support of land-grant colleges. Yet, the passage of the first Morrill Act was not an easy one: "The forces of greed, selfish expectations, blind opposition, myopic conservation, and blatant ignorance fortified by the weight of apathy of the general public maintained a standoff with the idealistic effects of a handful of dedicated visionaries."[3] Earlier critics of land-grant colleges included the private, elitist universities of Princeton, Yale, and Harvard. Charles Eliot, president of Harvard, declared that subsidies to agricultural schools were symptomatic of a "deep-seated disease" resulting from government interference in the affairs of citizens. Yet Harvard College had previously been the beneficiary of considerable amounts of public monies.

After ten years of debate and a veto by President Buchanan, midwestern farmers finally won passage of the Morrill Act. Although the Midwest with its family farms was the most democratic section of the country at the time, it was "the democracy of the small capitalist versus the Eastern aristocracy and the plantation aristocracy in the South."[4] The land-grant colleges were intended to provide instruction related to the practical needs of agriculturalists. As 80 percent of the 1860 population was rural and 97 percent of all farms were under 260 acres, the initial benefits from the land-grant colleges were meant to be egalitarian—but only for white males. The first Morrill Act benefited neither blacks, because 90 percent were slaves, nor women, who were barred from colleges in general.

U.S. Department of Agriculture

In 1862, three major statutory foundations of federal agricultural programs were created at a time when southern delegates to Congress were absent because of the Civil War. In addition to passing the Morrill Act—granting land to states to found and support agricultural colleges—and the Homestead Act—opening up the public domain to owner-operators—Congress established the Department of Agriculture. The USDA was the first client-oriented department in the federal government; it was exclusively concerned with the affairs of farmers. Its charter was comprehensive: to acquire and to diffuse among farmers useful information on subjects connected with agriculture "in the most general and comprehensive sense of the word" and to procure and distribute free, new, and valuable seeds and plants.[5]

The egalitarian purpose of the USDA was openly challenged by the second secretary of agriculture, J. Sterling Morton. His laissez-faire doctrine brought him into a head-on collision with Congress over the free distribution of seeds. In his 1894 report, this special agent for the purchase of seeds wrote, "In light of my experience as a former seedman, however, I consider the free distribution of seeds by this Department as an infringement upon and interference

Figure 10.1. Government and agriculture. (Top) U.S. Department of Agriculture Buildings, Washington, D.C., 1936. The USDA was the hope for family farmers, but it quickly sided with agribusiness. Photo by M. Kennedy, USDA. (Bottom) USDA Dairy Farm at the Beltsville Center in Maryland, 1942. USDA photo.

with a legitimate business, and I believe it should be abolished."[6] Despite Morton's attempt to terminate the free distribution of seeds, Congress passed a joint resolution to overrule him (but business interests triumphed eventually because the USDA no longer distributes free seeds).

The class interests in this example are clear. As long as economic practices such as the seed distribution are not profitable, the state provides these services, but once these activities become profitable, business interests want to take them over and usually the state lets them. By improving agriculture the USDA created a profitable private market for agricultural products and services. In this way, the State subsidizes agribusiness in the name of free enterprise, and taxes from the many enhance the private wealth of a few. This same process reappeared in the agricultural experiment stations. Largely publicly financed agricultural research in land-grant colleges has produced products and techniques that are marketed by private firms for private gain. The Hatch Act facilitated this development.

Hatch Act of 1887

The Hatch Act authorized federal funds for direct payments to each state that would establish an agricultural experiment station in connection with its land-grant college. The mandate was quite specific:

> to conduct original and other researches, investigations, and experiments bearing directly on and contributing to the establishment and maintenance of a permanent and effective agricultural industry of the United States, including *research basic to the problems of agriculture in its broadest aspects, and such investigations as have for their purpose the development and improvement of the rural home and rural life and the maximum contribution by agriculture to the welfare of the consumer*, having due regard to the varying conditions and needs of the respective states.[7]

Advocates of the act believed that the benefits of agricultural research should be enjoyed by *all* farmers. Since farming was not dominated by agribusiness in the 1880s, farmers needed the assistance of the government to do research on their behalf.[8] To assure widespread benefits from the agricultural stations, Section 4 of the act states that research bulletins be published at least once every three months and that free copies be sent to each newspaper in the state and, if requested, to farmers also. This practice continues today.

Morrill Act of 1890

With the expansion of the land-grant colleges, additional operating money was needed. The second Morrill Act of 1890 provided annual federal appropriations for the states to use. At the same time, Congress added a "separate but equal" provision authorizing the establishment of colleges for blacks. (Seventeen southern and border states took advantage of the act, which is discussed in greater detail later in this chapter.)

Smith-Lever Act of 1914

Out of the land-grant colleges and the experiment stations grew the need
to extend the teachings of the colleges and the research of the stations directly
into rural areas for the benefit of people there. The purpose of the extension
movement was to help rural people solve their work, social, and community
problems. The extension service—the third part of the land-grant college
system—was added in 1914 with the Smith-Lever Act, which financed state
extension workers. Lever, one of the sponsoring legislators, made clear that
extension agents "must give leadership and direction along *all lines of rural
activities*—social, economic, and financial."[9] The intent of all three parts of
the land-grant college system was to provide education, research, and services
to working farm families and rural communities. However, the consequences
for farm families, seasonal agricultural workers, and rural people in general
have been devastating.

Biases and Consequences

The total funds for agricultural research and development, public and
private, are now approximately $1 billion per year. Between $600 and $750
million of public funds are spent by the land-grant college system each year.[10]
Because people in rural America, particularly working farmers, farm workers,
and small business people, do not have the private resources and training to
do agricultural research, public funding of the land-grant colleges has been
justified since its inception. Agribusiness firms such as Safeway, Del Monte,
Tenneco, and Ralston Purina are capable of doing their own research and
need not be subsidized by taxpayers. Yet, instead of focusing on the needs
of rural *people*, the land-grant college system serves the interests of *capital*
and *technology*—in short, agribusiness.

How does the land-grant college system spend its money? Who makes
these decisions? Who benefits? Who gets hurt?

The purpose of the extension service is to disseminate the results of agricultural and rural community research to rural people and to help them solve
the problems they face. But as land-grant colleges and experiment stations
have a bias toward large-scale producers, extension workers, not unexpectedly,
share this same bias. Extension service workers devoted 86 percent of their
resources to the wealthiest third of America's farmers. On a per capita basis,
wealthy farmers received fourteen times as much attention from the extension service as low-income farmers.[11] Out of the 15,000 extension workers,
approximately 500 work in the area of rural development; the rest assist to
increase production. Generally, the plight of agricultural workers—whether
migrants or farm owners—is considered the responsibility of some other
agency, such as the Department of Labor or the Department of Health,
Education, and Welfare.[12]

This bias is particularly troublesome because the quality of life in many
rural communities is dismal and urgently in need of extension services. In

Figure 10.2. Agribusiness technology. (Top) The common silo becomes a $35,000 Harvestor in the upside-down world of agribusiness, which displays its hardware at farm festivals such as this one in Iowa, 1972. The blue and white Harvestors are a good indicator of high farm indebtedness and are concentrated on the largest dairy farms. (Bottom) Large-scale farm machinery on display at the 1972 Farm Festival in Iowa. The tractor sold for $35,000, the discs for $8,000.

the early 1970s, for example, 46 percent of the nation's poor were located in nonmetropolitan areas; median family income of rural people was 27 percent less than that of urban families; almost 60 percent of the nation's substandard housing was located in rural areas; and rural education attainments lagged far behind that of urban populations.[13] But these pressing rural issues have received little attention from extension workers or from researchers who develop the knowledge disseminated by extension workers.

Agricultural research and education has been geared to "efficiency" of production, meaning profitability, rather than to equality of farm families. For example, in 1979 the USDA announced the development of a new strawberry that was "uniquely adapted to mechanical harvesting." The berry is named "Linn" after the inventor, a USDA scientist. USDA claimed the Linn strawberry "remains firm when ripe, has good color and flavor, and features a prominent stem that can be removed easily." Letting machines harvest the strawberries, USDA boasts, could reduce hand labor by 75 percent.[14] In this case taxpayers' money was used to create unemployment among the poorest rural population—migrant families. Furthermore, such "biological efficiency" research is concerned primarily with increased shelf-life and tolerance for machine handling, both facilitators of profitability. It is never concerned with nutritional quality, and most consumers would rate such "efficient" produce to be vastly inferior in color, flavor, and texture.

In 1974, the U.S. General Accounting Office demonstrated the present research orientation of the USDA by documenting total allocations of USDA, land-grant college, and agricultural experiment station scientific person-years (referred to as standard man-years or SMYs). Only 404.6 SMYs were allocated directly to research the quality of community life in rural America, or just 3.8 percent of the total SMY commitment (a reduction from 5 percent in 1969). By comparison, 771.3 SMYs were assigned to improve biological efficiency of field crops and 569 SMYs were used to improve efficiency in the marketing system. These data translate to just one community and life-quality research SMY for every 161,000 rural Americans, compared to one food and fiber research SMY for every 6,400 rural Americans.[15]

Jim Hightower's 1969 study of U.S. land-grant college research also shows the enormous disparity between research investment in capital-intensive technology versus research for human needs:

1. 1,129 SMYs on improving the biological efficiency of crops versus only 18 SMYs on improving rural income.
2. 68 SMYs on marketing firm and system efficiency versus 17 SMYs on causes and remedies of poverty among rural people.
3. 842 SMYs on control of insects, diseases, and weeds in crops versus 95 SMYs to keep food products free from toxic residues from agricultural chemicals created by other land-grant scientists.
4. 200 SMYs on ornamental shrubs, turf, and trees for the affluent versus 7 SMYs on rural housing—a necessity for the poor.[16]

In 1971, the University of California spent $23 million for "organized research in agriculture." Over 95 percent of this money went into technological research; if it were broken down project by project, probably only 1 or 2 percent of the money was actually used for human-social research. This means that agricultural research is geared to higher yields and greater capital substitution for labor. Much of this research, therefore, contributes to the reduction, displacement, and impoverishment of the rural population.

Funding plays an important role in rural research. Private foundations and government agencies have vested class interests that are reflected in the agribusiness research they fund. Scientists study research problems where the money is, and the research money is in technological innovations, not improving the lives of rural families. There are plenty of important social research questions, but they threaten the status quo of agribusiness and agrarian capitalism. Instead, production problems are studied to find ways to increase private profits.

Despite the social problems of rural America, every land-grant college and experiment station has research programs intended to solve the technical problems of harvesting crops by machines.[17] Agribusiness firms such as Union Carbide, the second largest chemical company in the United States, benefit from this kind of research. As F. Perry Wilson, chairman of the company, said, Carbide is "developing chemistry-based products to capitalize on the mechanization of agriculture."[18] Carbide's new innovation, seed tape—a ribbon of polyoxide plastic containing seeds—has sold well to large-scale farmers.

Two detailed case studies of mechanization illustrate the specific impacts of the research done by the land-grant college system. The tomato harvester has already industrialized tomato production in California. The tobacco harvester, on the other hand, has only just begun to change economic and social relations in the Southeast.

The Consequences of Mechanizing Tomato Production
in California

Tomato harvesters were developed quickly. The first 25 harvesters were used in California in 1961. By 1964, 75 were in use; a year later, 250. The number increased to 1,000 in 1967, when approximately 80 percent of the California acreage was harvested by machines.[19]

Several universities and private firms contributed to research and development of the tomato harvester. The University of California at Davis and Blackwelder Manufacturing Company of Rio Vista, California, were the two principal parties behind the invention. Universities spent about $1.2 million and private firms spent about $1.9 million—a total research expenditure of $3.1 million. Mechanical harvesting displaced roughly 91 person-hours per acre of tomatoes harvested and reduced labor costs by $5.41 to $7.47 per ton. The estimated labor cost reduction was from $42.6 to $58.8 million. Traditional economic analysis, in which the distributional effects of the costs of innovations are assumed to be zero, indicated that the rates of return from the tomato harvester were quite substantial. But since 40 percent of the research monies came from land-grant institutions and the labor savings occurred for large-scale producers, the benefits of the research went to agribusiness and the social costs of labor displacement were borne by agricultural migrants and the taxpayers who had to provide their social welfare costs. Andrew Schmitz and David Seckler, agricultural economists, concluded that since compensation has not actually been paid to the displaced farm families who harvested tomatoes, "it cannot be concluded that society as a whole has benefited from the tomato harvester."[20]

William Friedland and Amy Barton, rural sociologists, identified six specific social consequences from the mechanization of tomatoes.[21]

1. Production of processing tomatoes became concentrated in California. Research on machine-harvestable varieties of tomatoes produced not only fruit with tough skins and with the capability of relatively easy separation from the vine but also tomato plants that would produce larger volumes of fruit ready for harvest at one time. Because of the unusually homogeneous and predictable weather conditions of California, the new varieties of tomatoes are harvested by the once-over technique, where machines cut the entire plant. Thus these new varieties are particularly profitable.

2. Concentration has occurred among fewer and larger growers because of the increased specialization necessary to grow tomatoes efficiently under the new conditions of new varieties and machine harvesting. The new machines require sizable capital investments. The $25,000 price tag of the early machines began the process of concentration, and the latest machines can harvest approximately 250 acres per season. For growers to enter tomato production now means that a major commitment must be made in acreage and in capital. Many growers have been eliminated from tomato production as a result of these increased commitments. Production tonnages went from 3.2 million in 1962 to 4 million in 1973. In 1962 California had about 4,000 growers; by 1973 the number declined to 597. Within ten years, scientists at the University of California at Davis had forced 85 percent of the state's cannery tomato farmers out of business.

3. Tomato production shifted regionally, away from the Stockton area in San Joaquin County to the southern San Joaquin Valley. The primary factor responsible for the shift was the sizable land units necessary for tomato production under machine harvesting conditions. The new varieties were also better suited to the environmental conditions of west Fresno County. In addition, the construction of irrigation networks as a result of the California Western Project provided vast quantities of subsidized water to the new areas. By contrast, the Stockton area had a relatively large number of medium-sized farms, soil conditions that were less than optimal for the new varieties of tomatoes, and a shortage of large quantities of cheap water. The drastic result was that the number of tomato growers dropped from five hundred in San Joaquin County alone in the early 1960s to less than six hundred in the entire state by 1972.

4. The introduction of tomato harvesters and the associated concentration of tomato production in the lands of a few growers contributed significantly to the success of price bargaining. Formerly, the large number of small growers

and their inability to maintain close and immediate communication placed the growers at a disadvantage in dealing with the processors. The weakness of growers in bargaining for tomato contracts was the normal weakness of the single individual who negotiated with a small number of centralized agencies, in this case, canneries. Under the new circumstances, the California Tomato Growers Association undertook statewide bargaining with the processors in 1975 after signing up over 65 percent of its membership under price bargaining arrangements. The few large-scale growers are now in a stronger position to bargain for higher prices than the previously numerous small growers, who were in greater need of higher prices. In fact, the state's top four tomato processors control more than 80 percent of the California tomato industry.

5. The nature of the harvest labor force changed from male to female, from Mexican to American, from migratory to settled. Before the tomato harvester, tomatoes were harvested by braceros, males recruited from villages in Mexico. The "winos"—migratory single men, often alcoholics—supplied another source of exploitable labor particularly during the "flash-peak" harvest seasons. The improved working conditions provided by tomato harvesting machines attracted a new labor pool, previously unavailable since only the poorest were willing to do stoop labor. By 1975 approximately 80 percent of the harvest labor force was female. "Housewives" are an ideal labor pool for growers. First, they can manage to schedule their work times around the seasonal employment needs of the growers. Second, since they are often secondary wage-earners and used to providing unpaid labor in their homes, they accept low wages. Third, these women are available at short notice and for very low recruiting costs. Fourth, they immediately return to their jobs in the home, and hence do not represent a "drain" on the social services of the local community. In effect, the local family structure provides growers with a dependable, low-cost, docile, readily available, invisible, and temporary labor source. No wonder conservatives preach the survival of the family!

In addition to changing the nature of harvest labor, mechanization also diminished the size of the labor pool required. In total, public funds paid 93 percent of the $1.8 million annual cost of mechanization production projects at the University of California campuses. These projects will eliminate most of the 176,000 harvest-time jobs in the state.[22]

6. Technological sophistication in harvesting tomatoes has not improved primitive labor conditions. For farm workers, machines reproduce factory conditions; indeed, conditions often are hotter, dirtier, and more uncomfortable than in industrial plants. To most growers, workers are simply hands employed to perform seasonal jobs, to be dismissed at will, and to be paid the minimum required. In industrial employment workers are treated more humanely because union contracts and/or federal legislation require it and because better treatment of workers improves productivity. This more sophisticated level of exploiting workers has not yet reached agriculture, which is dependent on seasonal labor. The prevailing ethos of most growers is that workers are endured and the sooner they can be replaced by machines, the better. As one grower revealed, "The machine won't strike for more money

Figure 10.3. Women sorting canning tomatoes on a mobile harvesting machine in Yuba City, California. Although these machines increase productivity per worker, they result in lower yields per acre and waste food because not all tomatoes are picked by the mechanical harvesters.

and will work when we want it to."[23] This attitude, reminiscent of nineteenth-century capitalists, is supported by the land-grant colleges, whose educational and research services facilitate agribusiness exploitation and displacement of farm workers.

The University of California at Davis recently developed a photoelectric color sorter for the mechanical tomato harvesters. Priced at $50,000 and up, the electronic device uses infrared lights and color sensors to distinguish a green from a ripe cannery tomato. Jack Deets, a corporate executive of a sorter manufacturer, predicts that in the next five years every tomato harvester in California will be equipped with an electronic sorter. In 1976 the electronic eyes had already displaced some 5,000 California farm workers. Those lucky enough to find work were given wage cuts of twenty-five cents per hour and their employment period was cut from eight weeks to one or two weeks.[24]

The strong commitment to mechanization of agriculture by the University of California is partially reflected in the conflicting roles university administrators play. Chester McCorkle, vice-president of the University of California, and Edward W. Carter, a member of the university's board of regents, both sit on the board of directors of Del Monte, a major beneficiary of mechanization research. Carter is also the managing partner of ASA Farms, the corporate owner of a million-dollar tomato cropland in Yolo County.[25] Through connections like McCorkle and Carter, private funding agencies—farm associations, canners, and chemical companies—determine the kind of research conducted at the land-grant colleges.[26]

Pending Consequences of Mechanization in the Flue-Cured Tobacco Industry of the Southeast

During 1967, about 295 million person-hours of labor were required to produce the nation's flue-cured tobacco crop. If currently available tobacco harvesters are adopted, labor input in tobacco production will be cut in half and competitive viability restricted to affluent farmers. The mechanical system requires a capital outlay of $52,000: $40,000 for bulk-curing barns and $12,000 for a harvester and support equipment. Operated at capacity (about 40 acres) the mechanical harvester is considered the least costly form of harvesting tobacco when wage rates exceed $1.35 per hour.[27]

The high equipment costs and the small size of production units, resulting largely from the restrictive federal tobacco acreage allotment programs, have served as an effective deterrent to extensive mechanization—so far.[28] In the Coastal Plain of North Carolina, the area with the largest allotments, the average production unit was only 8.9 acres in 1968. Farms having two or more sharecroppers averaged 19 acres of tobacco. The Piedmont area had even smaller production units, about 5.8 acres of tobacco per farm.

Tobacco is associated with special characteristics. The crop is produced primarily in the southeastern states. Flue-cured areas of North Carolina and adjacent counties in South Carolina constitute some of the more heavily populated rural areas in the United States. Much of this population relies on full- or part-time employment in the tobacco industry.[29] Mechanical harvesting increases the demand for skilled workers, who displace the traditional and largely female and child labor force. Family incomes of tobacco growers will decline as mechanization increases. From 1961 to 1971 technological change and mechanization on farms had already forced 300,000 persons out of the tobacco industry and tobacco-growing regions of the South—177,000 from North Carolina alone.[30] Consolidation of tobacco production to take advantage of economies of size associated with mechanization will cause a greater displacement of tenants than of part owners and full owners. Because there are twice as many black tenants (67 percent) as white tenants, blacks will be displaced even more.[31]

Hired workers are very important in the present hand harvest of tobacco. An estimated 215,000 workers (200,000 casual and seasonal, 15,000 full time) are employed in the flue-cured tobacco region. Although precise information on their social and economic characteristics is lacking, U.S. census data indicate that hired workers have low amounts of formal schooling and are mainly black. Hired workers will be the first to be displaced by mechanization.[32] The absence of data on hired tobacco workers and the lack of information on potential welfare costs, job requirements of local industries, and cost of retaining displaced workers reflect the research bias of the land-grant colleges in the Southeast. Yet the development of labor-saving technology has been and is being financed by federal funds at every experiment station in the tobacco states. Labor-reducing technology in tobacco is not restricted to mechanical harvesters; it also includes chemical growth regulators and weed control, mechanical transplanters, electronic temperature control and bulk curing, and mechanization in the marketing sector. In contrast, Robert McElroy found that not a single experiment station had a program to facilitate the adjustment of displaced agricultural workers. Except for some work by a small Economic Research Service group surveying mechanization of tobacco, no such efforts were occurring even at the federal level. McElroy concluded that "this is not a criticism of efforts devoted to the development of labor-saving technology; the benefits are sufficiently well-known. . . . In view . . . of the paucity of information on those displaced and the consequent and as yet nonquantifiable social costs, however, I do question the balance in the direction of national social efforts."[33]

The position of the land-grant colleges on tobacco research is illustrated by an exchange between Senator Adlai E. Stevenson III from Illinois and Dr. John T. Caldwell, chancellor of North Carolina State University, which had spent twenty-three years and over three-quarters of a million dollars to mechanize tobacco.

SENATOR STEVENSON. The average unit for tobacco production was 8.9 acres. The new technology requires 40 acres to be economically self-sufficient.

I understand that many small farmers cannot use this new machinery, and that they are going to get thrown out of business. As far as I can tell, nobody is addressing any attention to the problems of the 150,000 that are losing jobs.

You go beyond that and say this is a problem in our economy at large, and you could not be more right, the problem of technological displacement exists in industry as well as in agriculture, and, in both cases, we have not faced up to it in this country.

You are recognizing, I think, that we ought to face up to it, and I am suggesting that the land grant colleges ought to face up to it.

DR. CALDWELL. Senator, we do face up to it.

Let me get into the logic of what is being said. There is very little logic to the analysis being drawn. Earlier in the day, there was some suggestion by someone, perhaps by your questions, that it might be more economical to have hand labor and to exclude mechanized labor.

Every farmer makes his own decision in this business. If he wants to mechanize, he can mechanize [emphasis added].

If he does not want to mechanize, he does not have to mechanize.

He can either hire the labor that is available, if it is available, work a little harder himself, and keep his children in school if he wants to, or he can make a third decision, and he can get out of farming.

Nobody forces these decisions on him, but even this small tobacco farmer with the 8 acres or 9 acres can join with other tobacco farmers in some of the mechanization equipment he will use.

We help farmers with these decisions. We have all kinds of analysis that will indicate to a small farmer when he ought to mechanize, and when he should not.

If they use this information, fine, we make it available to them, we talk to them about it, we discuss it in field days.

It is fact that we are promoting to do away with labor. We help them make their own decisions as to whether they are going to buy a tractor or buy a harvester, or keep on doing it by hand, and his own family circumstances determine it. If his children say, "By gosh, Dad, I am not going to harvest tobacco anymore," and they go to town, there is nothing he can do about it, and the land grant colleges can not make him do it.

SENATOR STEVENSON. You said they have a choice. Most of them do not have a choice.

DR. CALDWELL. *They have a choice. They can go out of farming.* This has always been true. Newspapers can go out of business, too. The corner grocery can go out of business. *There is no way in the world for us to control the normal operation of the free economic system, where capital and labor and management goes where it wants to, and makes its own decisions* [emphasis added].

SENATOR STEVENSON. What I am trying to suggest is that this system is not free enterprise, that is why that farmer loses out, and it is only because of mechanization, it is because he does not have access to credit with which he can invest $52,000 for a machine to harvest his tobacco.[34]

The land-grant college system uses public funds to facilitate mechanization of agriculture, which will increase productivity and liberate farm workers from dangerous, back-breaking, and laborious work. Unfortunately, under capitalism the benefits of mechanization are not enjoyed by the people displaced—migrant workers and family farmers—but by large-scale producers and agribusiness. If the benefits of mechanization were used to retrain the affected people for alternative and higher-paying jobs and to lower food costs for consumers, increased mechanization would be a desirable public policy. But under the present circumstances, the benefits of agricultural technology are being siphoned off by agribusiness at the expense of the many rural people and taxpayers, on whose behalf the technological advances are justified as a rational national policy.

White Laws and Black Land-Grant Colleges

The bias of the land-grant college system has been and continues to be particularly devastating for rural blacks in the South. Traditionally, blacks have faced two distinctive handicaps compared to whites. First, they have

Figure 10.4. Mechanization: each cotton harvester picks as much in one hour as a person could pick in 72 hours.

represented a larger proportion of tenants and small-scale producers. Second, they have greater difficulty securing farm loans from rural banks in the South, in part because of discriminatory loan practices. The USDA and the land-grant college system in the South have added to these persistent problems.

When Congress passed the Morrill Act of 1890, primarily to increase appropriations to the already established white land-grant colleges, it also included the provision to establish colleges for blacks. These institutions were intended to serve as counterparts to the 1862 colleges, established by the first Morrill Act, in the states that insisted on a separation of races.

The act contained a "separate but equal" clause: "No money shall be paid . . . for the support and maintenance of a college where a distinction of race or color is made in the admission of students, but, the establishment and maintenance of such colleges separately for white and colored students shall be held to a compliance . . . if the funds received in such State or Territory be equitably divided."[35] The act specified that an "equitable" division meant equal or proportional: "The funds to be received under this act, between one college for white students and one institution for colored students . . . shall be divided into two parts. . . . *Institutions for colored students shall be entitled to the benefits of this act . . . as much as it would have been if it had been included under the act of eighteen hundred and sixty-two"* [italics added].[36]

Even if federal funds had been appropriated in proportion to the number of students in white schools, the 1890 colleges were at a disadvantage. No portion of the appropriations from the second Morrill Act could be used for the construction of new campuses. Only already established schools could be designated as 1890 colleges or the states themselves could pay for the construction of new colleges. In contrast, the 1862 colleges had received sizable land grants from the federal government to establish their campuses. Sixteen black land-grant colleges remain today (see Figure 10.5).

Black land-grant institutions have never been accepted by white land-grant colleges as equal members. Funding for black schools has violated the intent and the spirit of the Morrill Act of 1890. William Payne, program analyst with the U.S. Commission on Civil Rights, found that in 1968 the USDA "gave nearly $60 million to the predominantly white land-grant colleges, *150 times* the figure of less than $400,000 it gave to the 16 predominantly Negro land-grant colleges in the same states."[37] For example, in 1968 the federal government gave $5.8 million to Clemson University and $490,000 to predominantly black South Carolina State—almost *twelve times* as much federal money for three times the enrollment. The white University of Georgia, with ten times the enrollment of black Fort Valley, received nearly *twenty-four times* as much federal aid.[38]

According to the National Science Foundation, in 1970 the USDA provided approximately one-third of the federal funds received by the 1862 and 1890 institutions in the dual system states. Yet of the $76.8 million in USDA funds

Figure 10.5. Black and white land-grant colleges and universities. West Virginia discontinued its black land-grant college status of West Virginia State University in 1957.

allocated to these schools, almost 99.5 percent went to the sixteen white land-grant colleges; the 1890 colleges received a grand total of $383,000, or 0.5 percent. In short, the white institutions received *two hundred times* what the black institutions received. Even on a per capita basis, the white schools received about *thirty-five times* what the black schools received.[39]

In 1972 Congress finally took a first step to partially rectify this inequality by increasing the USDA appropriations for the sixteen 1890 colleges from the $4 million requested to more than $11.5 million. However, in the same year the sixteen white colleges in these same states received from the USDA approximately $94 million—almost $82.5 million more than the 1890 land-grant colleges.[40]

Both the Hatch and Smith-Lever acts directed state legislatures in the sixteen states with 1862 and 1890 land-grant colleges to designate "the college or colleges" that will receive federal funds. In all sixteen states 1862 colleges have always been chosen. From 1957 to 1971 black colleges received *nothing* from the Hatch and Smith-Lever acts funds.[41] In 1971 alone, the USDA, through these two acts, allocated approximately $87 million to white land-grant colleges in states with black colleges. In the same year, the main source of federal agriculture funding to the predominantly black colleges in these states (Public Law 89-106) allocated grants totaling $286,000. The white schools received approximately three times more from *two* agriculture programs than black colleges received from *all major federal programs combined.* The allocation to individual black colleges ranged from $22,000 to $12,000; the average was $17,687 per black institution![42]

The Cooperative Extension Service of the USDA was established by the Smith-Lever Act of 1914. The act says that where a state has two land-grant colleges, the appropriations for extension work "shall be administered by such college or colleges as the state legislature . . . may direct." Senator Smith of Georgia, author of the act, made the racist purpose of this provision quite clear: "We do not . . . want the fund if it goes to any but the white colleges."[43] His purpose has been served. In each of the states with white and black land-grant colleges, state legislatures have directed *all* federal extension funds to the 1862 colleges and the USDA has complied.

In addition to the institutional racism in the funding of extension service at black colleges, the extension service at white land-grant colleges in the South discriminates against rural blacks. The Cooperative Extension Service was totally segregated until 1964 and has done little since then to help the poorest and black farmers improve their standard of living. Consequently, a 1970 study found that 80 percent of black farmers in Alabama, for example, had never been visited by an extension agent.[44] The office of the inspector general of the USDA found consistently that extension services in the Southern and border states discriminated persistently from 1965 to 1972 in such matters as the holding of segregated meetings, the use of segregated mailing lists, the refusal to permit white agents to serve blacks or black agents to serve whites, and discriminatory hiring and promotion policies.[45]

As a result of these racist practices, black farm operators have found survival in southern agriculture more difficult than whites have. They began the period of technological advance on much smaller farms—45 acres as opposed to more than 120 acres for whites in 1920—and have been at a serious disadvantage ever since, for economic as well as racist reasons, in trying to enlarge their farms.[46] Although the average size of black farms in the South almost doubled between 1964 and 1974, the average black farm remains under 100 acres, well below the average size of a white farm in 1954.

According to Virgil L. Christian and Carl C. Erwin, the rapid demise of black family farmers—like family farmers in general—may be due more to rapid technological changes, all of which increased the minimum size of viable farms, than to economies of size that were premised on fixed technology. The research on the mechanization of various crops by the land-grant colleges was primarily responsible for rapid technological changes and thus responsible for the rapid decline in the South of the smaller-sized black family farmers and of tenant farmers in particular.[47]

Public money in the land-grant college system gravitates upward to those who are well off rather than downward to those who are poor. Hence, as the 1862 colleges spend disproportionately more public research funds for prosperous agribusiness, the 1890 colleges receive disproportionately less money to serve low-income, small-scale, rural black families. The unequal funding of black colleges is particularly ironic because these schools serve the very people the land-grant college system was established for—the plain people, to use Lincoln's phrase. Sixty percent of the nation's blacks still live in states with 1890 colleges, and close to 50 percent of these blacks reside in rural areas. The clientele of the black land-grant colleges have an average income of less than $5,000 per year. The approximately 50,000 students that are enrolled in the 1890 colleges come mostly from low-income families.[48] In addition, most of the funds at the 1890 colleges are used to aid rural people rather than to displace farmers, as the white land-grant colleges do. For example, in 1972 the black land-grant colleges received $12.8 million for research and extension from the USDA; most of these funds were used to assist people to improve their incomes, welfare, health, and communities.[49]

Doubly ironic is the fact that the very agency (USDA) overseeing the unequal allocation of land-grant college appropriations praises black colleges for having "rapport" and "unique channels of communication with the unreached and hard to reach" in rural areas of the South. The USDA also admits that with the slightly increased research funding the black land-grant colleges have recently received, they have already begun to carry out innovative programs, particularly aimed at helping the rural poor.[50]

The discriminatory funding and operational practices of the land-grant college system have persisted despite the passage of the 1964 Civil Rights Act. This act sought to put the enormous potential political clout of federal agencies and programs at the service of racial justice. Title VI of the act gives minorities equal protection, access, and benefits from federally financed programs. Yet the USDA claims that this act does not apply to the funds from the Hatch and Smith-Lever acts because the benefits conferred by the white agricultural extension and research services are available to all persons. This same argument is used by the USDA to justify the high priority for research on the mechanization of agriculture throughout the land-grant college system as a whole. But in fact, the allocation of public funds and practices of the 1862 colleges have benefited almost exclusively white farmers and particularly large-scale white farmers.[51]

Who Benefits and Who Pays?

The tax-supported land-grant college system has used its vast financial, research, and extension resources for the benefits primarily of large-scale producers and agribusiness firms. These are the farm operations with the acreage and borrowing power to invest in the new machinery and technologies. Capital-intensive producers, like food-processing corporations and vertically integrated input industries, have the financial resources and managerial skills to mechanize the food system. These farmers hire the vast majority of migrant and permanent farm workers and therefore have an economic incentive to reduce their labor bill and potential labor problems through mechanization.[52] At best, land-grant college research benefits are geared to the needs of no more than 35 percent of the country's farmers, those with annual sales in 1974 above $20,000. More realistically, mechanization research is designed to meet the specifications of the largest producers: the 7 percent of all U.S. farms in 1974 with annual sales of $100,000 or more that produced 54 percent of the total value of agricultural sales.

Taxpayers, through the land-grant college system, have given large-scale farmers technology suited to their scale of operation and designed to increase their profits. While mechanization research is not designed to force the majority of American farmers off farms, it has that effect. Farm workers and most farmers are the victims of mechanization, in part because they are not involved in the design of agricultural research for their needs. Severe neglect and discrimination against the rural poor by agricultural researchers and extension workers, and indifference to the impact of change upon small-town people, are thoroughly documented by presidential commissions, the U.S. Civil Rights Commission, congressional committees, and scholarly studies. Jim Hightower, director of the agribusiness accountability project, concluded that "the experiment stations at land grant colleges exist today as tax-paid clinics for agribusiness corporations, while others who need publicly supported research either benefit only incidentally, are not served at all, or actually are being harmed by land grant research."[53]

Part 5

Rural Consequences of Agribusiness

11
The Plight of Seasonal
Farm Families

The plight of seasonal and migrant farm workers is an expression and a consequence of agribusiness. These farm workers' lives represent the most visible form of human degradation in rural America. In its most developed form, migratory labor is a proletarian class, dependent on agrarian capitalists and their machines. The existence and importance of seasonal farm labor depends not so much on agri*culture*, a way of life, as on agri*business*, a way of making money. Because this labor pool is created by and for large economic interests, seasonal migratory workers are the conscious and unconscious weapon of agribusiness interests that threaten to destroy family farms. The exploitation of seasonal farm workers on large-scale farms leads to their poverty, which facilitates the prosperity of agribusiness and, hence, contributes further to the demise of family farms.

Family farms are based on family labor with occasional hired workers. In 1964, half of all farmers hired no labor, and another 43 percent of farmers hired less than one person-year of labor. Based on the labor requirement definition of family farms in Chapter 2, 93 percent of farms were family farms and the remaining 7 percent were large-scale farms. In 1977 farmers with sales of $200,000 or more accounted for 6 percent of all farmers who hired agricultural labor, yet these farmers paid 56 percent of the total dollar value of such help. Larger-than-family and industrial farmers employ almost all seasonal and migratory labor. Measures of efficiency on these farm units must take into account how these workers are treated and how much they are paid.

On family farms, labor relations are immersed in social relations. Under agrarian capitalism labor relations result in labor conflicts, but on family farms the severity of these conflicts emerges only as family disputes. The personal and labor relations of family members become confused, which benefits farm owners and ultimately agribusiness. The losers are those on family farms who do not own resources—wives, children, and hired farm workers.[1]

Input and output agribusiness firms squeeze both family farmers and large-scale producers, who in turn seek to minimize their labor expenses—the only production factor over which they have any control. On family farms, this means paying little or nothing to children and wives. In structural terms, husbands represent capitalists, their wives and children the proletariat. On large-scale farms, holding down labor expenses means paying migrants the lowest wages and providing them with the poorest housing possible. Farm owners and operators, regardless of their scale of production, have little choice but to exploit family and seasonal labor in a market economy that constantly threatens farmers with bankruptcy. Decisions in the marketplace and in government create the conditions to which family farmers and seasonal farm workers respond. Ironically, small farmers, even though they are resource owners, are in a situation similar to migrants', relative to large-scale producers. As one farmer expressed it: "We're both small fry. The bigger you get, the less is known about the way you make your money—and keep it, the tax laws see to that. If you get big enough, you even get to be a trustee of a college."[2]

With unstable residence, low income, and no control over land, migrant and seasonal workers have little influence within their communities on the wages paid and the legislative protection received. The decline in people's access to land, a consequence of unabated farm enlargement and concentration of farmland ownership, is a critical factor in shaping the problems of farm workers as well as family farmers and merchants in small towns.

In congressional committee hearings liberals have recognized the special plight of seasonal farm workers. Yet their legislative programs to help migrants have been few and piecemeal and have focused on the victims of large-scale farming rather than on the sources of farm problems. By contrast, conservatives blame the low standard of living of migrants on racial and/or personal characteristics. As one North Carolina grower said, "Do the 'do-gooders' know that half of the migrants are alcoholics and the other half of them are running from the law? A lot are mental incompetents. They have different standards than we do."[3]

Agribusiness claims to be proud to provide jobs to migrants, but when workers demand higher wages or organize themselves in any way crops are quickly mechanized. When cost/profit ratios allow, growers prefer to replace bothersome field hands with machines. USDA policies and land-grant college research greatly facilitate this mechanization. Replacing seasonal labor with machines has the double advantage for agribusiness of eliminating the "migrant problem" and of concentrating economic power even more in the hands of large-scale producers. Seasonal farm workers and their dependents, as human beings, are totally discounted by agribusiness—as they must be for agrarian capitalism to function and survive. Migrants represent the most extreme form of direct exploitation; they are treated as commodities, not complete people with pride, feelings, and needs.

Historical Background

Seasonal farm work has always been associated with industrialized agriculture.[4] In the 1870s large farms in the Midwest and California began specializing in wheat production and hiring itinerant workers for the harvest. While the bonanza farms of the Red River Valley were hiring newly arrived immigrant settlers, who could not make a living on their own farms, California growers were employing racial minorities. In California, Native Americans were the first farm workers and the Chinese were the first immigrant group to work in the fields. The Chinese were originally brought to the state to work on the transcontinental railroads; thereafter, they became a large new labor pool for the newly developed fruit and vegetable farms. As demand for labor mounted, more Chinese workers were imported. Their numbers increased in California from over 10,000 in 1869 to more than 130,000 in 1882.[5] With the supply of Chinese workers drying up because Congress prohibited their contract labor in 1885, a new supply of agricultural labor became necessary. By 1910, California had over 40,000 Japanese workers, most of whom were young single males working in agriculture, but soon exclusionary, racist laws were again passed "to stem the tide of the yellow peril."

Mexicans became the next major source of cheap agricultural workers. By 1920, nearly 100,000 Mexican nationals were in California. With the exception of the 1930s, Mexicans, Mexican-Americans (Mexican born), and Chicanos (American born) have provided the bulk of the farm labor in California and border states for sixty years. During the Dust Bowl years of the 1930s, 300,000 small-scale farm families arrived in California from Oklahoma, Arkansas, Missouri, and the Texas panhandle. Foreclosures by banks and inadequate federal relief programs displaced farm families from their drought-ridden land. Independent farm families, who had never worked for wages, had no alternative but to become day labor in agriculture. The ideal of the family farmer became the reality of the proletarian field worker.

At various times, California growers also used other Third World people such as Filipinos, East Indians, Arabians, Jamaicans, and people from U.S. dependencies. In addition, during both world wars, school children, boy scouts, prisoners, and homemakers worked in the fields.

The Bracero Program

The wartime demand for labor in industry threatened the supply of domestic agricultural labor. Growers received special dispensation (because the use of foreign contract labor was illegal) to use Mexican contract workers (braceros) during World War I and during and after World War II. Under Public Law 78 the bracero program (1942-1964) was the first governmentally administered system of migrant labor in agriculture. The federal government bore the costs of recruiting, transporting, feeding, and housing minorities. In 1945, U.S. taxpayers paid $21.6 million to place about 50,000 braceros at the disposal of growers—an average of nearly $450 per bracero—more than the average Mexican farm worker was able to earn in wages![6]

By law, domestic workers had priority in applying for and claiming work for which bracero labor was officially authorized. Wages for braceros were fixed at the level "prevailing" for domestic workers. To assure that the law was upheld, payroll records were required to be open to workers and to the communities into which the braceros came. But the USDA, state farm placement services, and growers' associations denied workers access to the records.[7]

Figure 11.1. Bankers foreclosed on farm families with inadequate incomes during the Depression, which led to their flight to the West Coast in the 1930s. FSA photo.

Contrary to law, wage rates and working conditions deteriorated as the number of braceros increased.[8] These results were not surprising given the number of officials assigned to enforce Public Law 78: the entire state of California had not more than a dozen on duty in the fields at any one time. One officer was responsible for the entire Imperial Valley, another for the Salinas Valley, another for Santa Clara. Without secretaries, these agents spent much of their time on paper work.[9]

Public Law 78, in its concept and especially in its practice, expressed the strong alliance between agribusiness and government agencies. At its peak, nearly 500,000 braceros were imported into the United States in a single year, nearly 100,000 into California alone.[10] Substantial public subsidies were thus used to assure an abundant, cheap foreign labor supply to agribusiness. The bracero program also provided a rhetorical shield behind which growers could hide their advantages and abuses.[11]

In 1964, against the militant opposition of agricultural employers, the combined efforts of church and nationalistic groups as well as organized labor succeeded in terminating the bracero program. The massive flow of cheap labor from Mexico appeared to be at an end; but the decrease in legal entries brought about by the end of the bracero program and a more restrictive immigration policy were generally compensated for by a sharp rise in illegal migrants or "wetbacks." This was not surprising because the problems of underdevelopment in Mexico and the need for a cheap and docile labor supply in the United States had not changed; only the laws separating the countries had.

Undocumented Workers

Between six and twelve million foreigners illegally immigrated to the United States between 1968 and 1978. Nearly half worked in agriculture. The standard view is that undocumented immigrants (or illegal aliens as they are more commonly called) pose a threat to the U.S. economy by taking jobs and utilizing social services at the expense of U.S. taxpayers. This mythology is based on the erroneous assumption that the U.S. economy has only so many jobs to be filled. The opposite is actually true: the economy is elastic, and allowing more people to work produces more wealth for all. Undocumented immigrants, as producers and consumers, actually help the economy. The chief administrative officer of Los Angeles County showed that in fiscal year 1975 undocumented aliens contributed $171 million in federal and state income taxes. The San Diego Human Relations Commission found that undocumented immigrants earned an estimated $260 million in wages annually in the county. They also spent $150 million in the county, and paid $48 million a year in state and federal taxes as well as sales taxes. The Linton report, commissioned by the U.S. Department of Labor in 1975, concluded that the average wage of undocumented workers in border counties was $1.74, producing an annual income too poor to be taxed. Nevertheless, 73 percent of them had federal income taxes withheld and 77 percent had social security payments deducted. Only 31.5 percent filed U.S. income tax forms, though most would have been eligible for refunds.[12]

Undocumented farm workers in Arizona, who won a contract guaranteeing them, for the first time, decent living and working conditions, expressed the moral basis for any discussion on "illegal aliens": "The law of God says that we are all equal; it is the law of man [capitalists and their supporters] that separates us. As brothers we are all equal. Whatever the color or race, we all have the same feelings."[13] As workers, most Americans need not concern themselves with whether workers are legal or illegal but whether all workers in this country have all of the available protection they are entitled to. Instead, the national emphasis is on keeping undocumented workers and especially Mexicans out of the country. Foreign migrants, legal and illegal, do not initiate migration; employers do by recruiting migrants for jobs in the secondary market at wages considered unacceptable by native workers.

In fact, the U.S. border is used to externalize the social and economic costs of underdevelopment. The process of underdevelopment in Mexico, starting with Spanish colonial rule and continuing today with U.S.-based multinational corporations, results in the displacement of people from rural areas and the absence of employment in cities to absorb them. The use of legal and illegal foreigners works to the advantage of elites and to the disadvantage of farm workers in the United States and Mexico. The structural problems of underdevelopment confront the Mexican elites with massive unemployment (estimated to be about 20 percent) and a potentially explosive lower class. Immigration to the United States provides a safety valve by alleviating the tensions and costs of income maldistribution and thus allows Mexican elites to export some of the social costs of foreign and internal capitalist development.

In the United States, Mexican immigrants provide employers with several advantages.[14] These workers represent a growing reserve army of labor that forces unskilled wages down by providing a threat of easy replacement to domestic workers. Undocumented Mexicans represent, above all, cheap, willing, and available labor power. They have no rights and are not subject to the usual legal restrictions affecting nationals concerning fringe benefits and occupational protection of workers. The constant threat of discovery—particularly exposure by employers—makes them willing to work at much lower wages.[15] These conditions are advantageous for the wealthy but compound the problem of persistent rural poverty in the United States.

Growers are assured of low-wage migrants because the world is organized as a center-periphery economy. The United States represents a center country that collects raw materials, talent, and migrants from peripheral countries. In the world space economy, Mexico represents a peripheral country, dependent on center countries for its prosperity. In both the center and periphery, workers are on the margin of full participation in society; unable to compete with the elites in their respective countries, they compete with each other.[16] That is, braceros and now undocumented workers from Mexico have competed with U.S. migrants while the agricultural elites in each country benefit from the center-periphery economic linkages. The U.S. government, in the form of the Immigration and Naturalization Service, serves the interests of U.S.-based multinationals, who created the structural problems that generate Mexican migration to the United States in the first place and who also benefit from this cheap labor while it is in the United States.

Agribusiness exploits migrants' linguistic and cultural distinctiveness, their ignorance of common American practices, and their hunger—sometimes employing threats of punishment—to create an adequate labor pool for agriculture. The conservative ideology of blaming the "coolie," the "Jap," or the "brown face" for their migrant status has been lucrative to growers and has obscured the migrants' need to be protected like industrial workers. Growers see agricultural labor from their own economic perspective, in which the ideology of the upper classes is consumed as objective truth by the lower classes. With regard to migrant workers, the state actualizes this conservative capitalistic ideology based on the following assumptions:

1. Workers must accept low wages, long hours, hard work, poor housing and diet, and hazardous working conditions.
2. Workers must be available precisely when and where demanded by growers, in larger numbers than required, so that growers can pay less and have a cushion against a possible hot spell with quickly ripening crops.
3. Growers must be under no obligation to care for these workers before or after the harvests and must provide (substandard) housing only during harvests.
4. Regardless of how closely migrant working conditions resemble industrial work, growers must not be asked to meet the social obligations that apply by law to every other business in which a few people make profits from the labor of other people.
5. Workers must have no voice in the terms of their employment; unionization would destroy agribusiness.[17]

Similar assumptions were made by nineteenth-century industrialists, who changed their behavior only when workers organized themselves to demand a higher standard of living. Although monopoly capitalism now predominates at the national level, an earlier form—industrial capitalism—still exists in agriculture. Ironically, although today's seasonal farm workers work under the most industrialized form of agriculture, they represent a nineteenth-century proletarian class.

According to Karl Marx, capitalists *buy* the labor of workers with money; workers *sell* their labor for money. But this is merely the appearance. In reality what workers sell to capitalists is their *labor power:* "Labor power is a commodity, neither more nor less than sugar. The former is measured by the clock, the latter by the scales."[18] Treating humans as commodities allows the greatest possible extraction of surplus value or profit from workers.

Maintenance and Renewal of Migrant Labor

The basic advantage of migrant labor is the externalization of the costs of labor force renewal and low-wage labor. Foreign migrants—essentially Mexicans, but also Puerto Ricans and French Canadians—depend primarily on their own country and to a lesser extent on the United States for employment. Internal migrants—legal and undocumented workers from Mexico—depend solely on employment in the United States. Both foreign and domestic migrants occupy marginal occupations, such as farm labor, and have separate places for employment and for their home base. The maintenance and renewal of migrant labor take place in geographically separate locations. The costs of subsistence (maintenance) and of raising a family of future migrants (renewal) are transferred from the regions in which migrants work to the regions or countries of the migrants' origins. Thus one class and regional economy, for example, growers in Colorado, are pitted against another class and region, such as Chicanos from southern Texas. Center-periphery relations between the United States and Mexico and within the United States create uneven international and national development. The economic dependency of the peripheral regions on the centers creates the flow of migrant labor, which further strengthens the center and weakens the periphery.

Foreign migrants have traditionally been preferred by agribusiness because immigration laws could be used to separate workers from their families. The costs of labor force renewal (raising families) could thus be borne abroad while the U.S. government and local employers would be responsible for maintaining migrants only during their term of employment.

Although foreign migrants are controlled through immigration laws and work contracts, domestic migrants are controlled within class divisions along racial and resident status. When domestic migrants have joined together to organize, foreign migrants, mechanization, and/or violence have been used to suppress them. Ernesto Galarza, a specialist in Latin-American affairs, described how the bracero program fostered the replacement of domestic labor with Mexican labor paid at rates well below what domestic workers could survive on.[19] Wages for domestic workers reflected only maintenance costs; renewal costs were externalized to Mexico. The segmentation of migrants into racial, foreign, and domestic groups has obstructed the development of effective union organizations. The ability of domestic labor to organize itself is severely limited by the power of growers, who have gained monopolistic access to external labor supplies. Police suppression of labor protests, exclusionary federal labor legislation, immigration laws, and Supreme Court rulings favoring growers have allowed growers to achieve their monopolistic control over agricultural labor.[20]

Growers also manipulate the wage rates they pay. Theoretically, the prevailing rates are determined by market forces. In fact, they are established unilaterally by the growers according to the same criteria followed by the South African Chamber of Mines: "A wage is fair to one's neighbor in that it is no higher, and a wage is fair to oneself in that it is no lower."[21] Unilateral wage fixing is only one consequence of the concentration of landownership and the vertical integration within the cannery industry.[22]

The dominant ideology in the United States includes concepts of "justice" and "equality"; consequently, the U.S. government has frequently resisted the blatant exploitation of foreign agricultural workers. During the bracero program, growers were required to prove a shortage of domestic labor and to adhere to specific regulations regarding housing and minimum number of work days. However, agribusiness gets its way through lack of enforcement of these provisions and the absence of laws covering domestic migrant labor. The dominant ideology conceals the practices of growers and obscures the plight of migrant workers and their dependents. (Figure 11.2 shows the location of a labor camp outside the city of Imperial. The camp is "concealed" both physically and socially.)

Under the capitalist mode of production, migrants represent one segment of the agricultural labor force struggling to maintain itself from day to day at subsistence levels. Large-scale farmers represent the other extreme of the agricultural labor force, which devotes more time and money to labor processes. That is, prosperous farmers send their children to college, often to get master's and doctorate degrees in agriculture and related fields, to facilitate their long-term survival in the marketplace. Consequently, the gap between farmers, who often represent agribusiness, and seasonal agricultural workers grows.

Numbers of Seasonal Farm Workers

Federal data collection on seasonal and migrant farmworker families is deficient and biased toward growers and agribusiness. One of the most serious

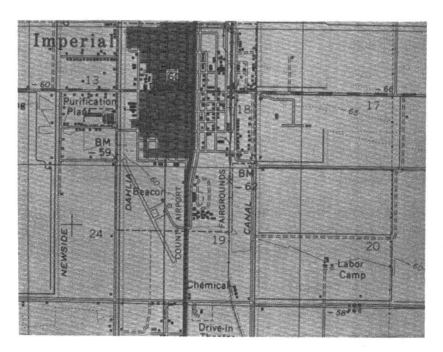

Figure 11.2. Topographic sheets, like this one from California's Imperial Valley, reveal the location of labor camps, even though growers prefer to keep them out of sight.

data problems is the gross underestimation of the farm-worker population, which is caused by the arbitrary and restrictive definitions of "farm work," "seasonal," and "migrancy." The effect, regardless of its intent, is to reduce the number of migrants and therefore the magnitude of the farm-worker problem for agribusiness, federal agencies, Congress, and the general public. Another serious deficiency in farm-worker data collection is the federal reliance on farm employers for the information rather than on farm workers themselves. Employers tend to report only the number of farm workers actually employed, and virtually no data are available on the living conditions of farm workers—their income, health, and employment needs. With such insufficient and biased data available to document the need for federal assistance to farm-worker families, progressive groups and legislators find it difficult to justify federal expenditures for a population that most federal data services have nearly counted out of existence.[23]

There are five separate data sources on farm workers, in addition to five special federal farm-worker programs. Agencies disagree on how many farm workers exist and on how to define them. Table 11.1 shows that each data source, using different definitions, provides different estimates of the number of seasonal and migrant farm workers. The variation ranged from a low of

TABLE 11.1
Five Estimates of Seasonal and Migrant Farm Workers

Data Source	Hired Farm Workers		
	Seasonal	Migrant	Total
U.S. Department of Labor, Employment & Training Administration (1975)	727,922	169,420	897,342
U.S. Department of Agriculture, Statistical Reporting Service (1975)	no breakdown given		1,376,700
U.S. Department of Agriculture, Economic Research Service (1975)	875,000	188,000	1,063,000
U.S. Department of Commerce, Census of Population (1970)	no breakdown given		2,345,000
Rural America, Inc. (1976)	1,016,911	235,023	1,251,934
Dependents	2,133,533	413,875	2,547,408
Total	3,150,444	648,898	3,799,342

Source: Where Have All the Farm Workers Gone? Research Report No. 1 (Washington, D.C.: Rural America, 1977), pp. 4, 7, 9, 11b, 12, and 64. The Census of Agriculture also counts these workers but it "contain[s] so many duplications that federal statisticians consider the data meaningless (p. 11)." For example, the 1969 census reported 5,125,604 seasonal farm workers!

897,342 to a high of almost 4 million. Rural America, an independent rural research organization, found that their own estimates were 34 percent higher than the highest federal statistical source and more than six times higher than the lowest federal estimate. The combined worker/dependent estimate made by Rural America was 46 percent higher than the U.S. Health Service Administration's estimate of 2.6 million persons.[24]

For their estimates, Rural America used the definitions of the most conservative data source, the U.S. Department of Labor's Employment and Training Administration. *Seasonal* (nonmigrant) farm workers are defined as having worked in agriculture at least 25 days but less than 150 consecutive days at one firm during the preceding twelve months. This group also includes persons for whom agriculture is not their major occupation, such as students and homemakers, even if they worked more than 150 consecutive days in agriculture. *Migrant* farm workers are defined as those who did not return home in the same work day. This group also includes food-processing workers who had to travel and worked less than 150 days.[25] Rural America applied these definitions to employed and unemployed farm workers and food processors and included their respective dependents as well (see Table 11.2). These data show farm workers as humans who do not always find employment and who have spouses, children, and relatives to care for. When these concerns were

Figure 11.3. Factories in the fields. Seasonal workers harvest grapes for wine in Delano, California. The converter belt speeds up harvesting. The winery is in the background.

taken into consideration, the number of farm workers increased sharply. Dependents alone accounted for more than twice the number of all farm workers. Cannery workers share many of the same conditions with seasonal field workers, whose numbers increased the total seasonal agricultural work force in the United States to 1.6 million, with a dependent population of 3.3 million.

Although farm workers are employed in every state, most are found in five states—California, North Carolina, Texas, Florida, and Washington. These states accounted for 54 percent of all U.S. farm workers and their dependents in 1976 (see Table 11.3). The largest 31,000 farms (less than 1 percent of all farms) employed 41 percent of all farm workers and had 24 percent of all farm sales in 1969. Fewer than 13 percent of all farms hired 78 percent of all farm workers. The Agribusiness Accountability Project estimated that 75 to 85 percent of farm workers were employed by major food and industrial corporations, such as Dow Chemical, Heublein, Coca-Cola, Penn Central, Bank of America, Campbell Soup, Prudential Insurance, Atlantic Seaboard Railroad, Union Carbide, Goodyear Tire and Rubber, and Safeway. In Illinois, 80 percent of farm workers were employed by four large corporations.[26]

TABLE 11.2
Rural America, Inc.'s Estimates of Migrant and Seasonal
Farm Workers, Food Processors, and their Dependents, 1976

Employed Farm Workers	1,251,934
Their Dependents	2,547,408
	3,799,342
Unemployed Farm Workers	147,728
Their Dependents	300,575
	448,303
Sub-Total	
Farm Workers	1,399,662
Their Dependents	2,847,983
	4,247,645
Food Processors	200,000
Their Dependents	408,000
	608,000
Total	
Farm Workers & Food Processors	1,599,662
Their Dependents	3,255,983
	4,855,645

Source: Where Have All the Farm Workers Gone? Research Report No. 1 (Washington, D.C.: Rural America, 1977), p. 65.

Although small growers employ relatively few seasonal farm workers, they tend to pay them less and treat them worse than large growers do. They also tend to hire the greatest number of undocumented field workers. But small growers are being squeezed by canners and have little choice but to exploit their seasonal workers. It is not lack of humanity but rather an inhumane economic context that forces family farmers especially into such labor relations. Small growers seldom blame agribusiness for bringing about this economic context because they must operate under its conditions. Racism provides a convenient explanation by blaming the victims themselves, the migrants, for their working conditions. Thus the actions of small growers are justified by the ideology of their own adversaries in free enterprise while contradicting the agrarian ideal.

Migratory Streams

Migrant farm workers, who leave their homes to work in areas beyond a daily commuting range, follow three major routes, all of which originate in southern border states (see Figure 11.4). The main stream flows out of Texas, especially the Rio Grande Valley and particularly Hidalgo County, from which over 40,000 Mexican-Americans leave each spring to do farm work elsewhere in the United States.[27] Chicanos make up 80 percent of the population in the Rio Grande Valley. One-third of them are farm workers who are forced to migrate: those who did find work in the valley in 1977 averaged twenty hours a week at $1.30 per hour. The valley has the nation's greatest concentration of poverty, with half the families, all Chicanos, earning annual incomes below the poverty level. The people call it the Valley of Tears.[28] The absence of jobs and social services forces farm workers to migrate

TABLE 11.3
Top Five States with Migrant and Seasonal Farm Workers and
Their Dependents, 1976

	Workers and their Dependents		
State	Migrant	Seasonal	Total
California	156,496	530,501	686,997
North Carolina	44,145	600,538	644,683
Texas	26,887	500,514	527,401
Florida	45,298	213,831	259,129
Washington	37,085	134,401	171,486
	309,911	1,979,785	2,288,996

Source: *Where Have All the Farm Workers Gone?* Research Report No. 1 (Washington, D.C.: Rural America, 1977), p. 67.

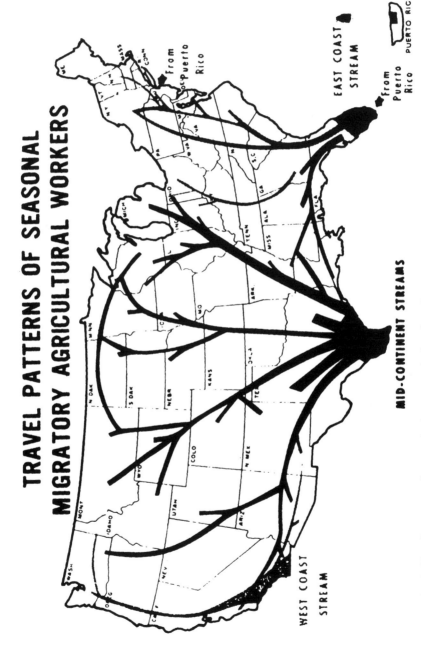

Figure 11.4. Travel patterns of seasonal migratory agricultural workers.

northward and westward covering most of the western, northern, and mid-western states. For example, migrant families might go to northern Michigan to harvest cherries, and then to Indiana and Ohio for tomatoes before arriving in Florida for the October and November vegetable harvest. The major crops involved in this stream are fruits and vegetables, sugar beets, and cotton.

The second migratory stream starts in southern California. The itinerary of the workers, mainly Mexican-Americans, might include early May cherries in the Sacramento Valley; crops in Idaho and Montana in July and August; peaches and pears in Oregon and Washington during late August; and olives in California in late September and October.

A smaller migratory stream consists of workers from Florida and other southeastern states for the Florida citrus and winter vegetable harvest. Migrants work northward during the spring and summer through the Atlantic coast states, sometimes as far north as New England. Blacks from the inner cities of the eastern megalopolis often constitute a large proportion of the East-Coast stream.

Growers in Texas, California, and Florida employ intra- and interstate migratory farm workers. States with large numbers of out-of-state migrants have even larger numbers of within-state migrants. In California, which leads all other states in employment of migrants and seasonal farm workers, the average number of intrastate migrants is about 1.4 times the average number of interstate migrants, who total over 200,000 at the peak harvest season.

The principal crops using seasonal labor in the U.S. today are fruits and vegetables—especially those for the fresh market—along with tobacco and nuts. The largest number of seasonal workers harvests strawberries, tomatoes, beans, apples, grapes, citrus fruits, peaches, and cherries. Since California specializes in these labor-intensive crops and has relatively large farms, which use disproportionately more hired labor than smaller farms, the largest number of seasonal and migrant farm workers is found in that state (see Figure 11.5).

Child Labor: Family Togetherness or Exploitation?

Children of farm workers were left out of the first effective child-labor laws that were passed in 1938, when the Fair Labor Standards Acts (FLSA) placed a limit on the age children were allowed to work in business and industry. Not only were children working in agriculture left unprotected by the child-labor laws, but their parents were also excluded from critical labor legislation. This prevented them from seeking adequate wages that would allow them to take their children out of the fields, just as industrial workers had taken their children out of the mines and factories earlier. After decades of fighting to have agricultural child labor included under the FLSA, Congress passed an amendment in 1974 making it illegal for growers to employ children under 12; children between 12 and 16 may be hired after school hours or with a special permit. Despite the new federal amendment and the old state truancy laws, most farm workers admit that they work alongside children, who are there illegally.

For growers, the matter is quite simple. "Kids not working is for the birds,"

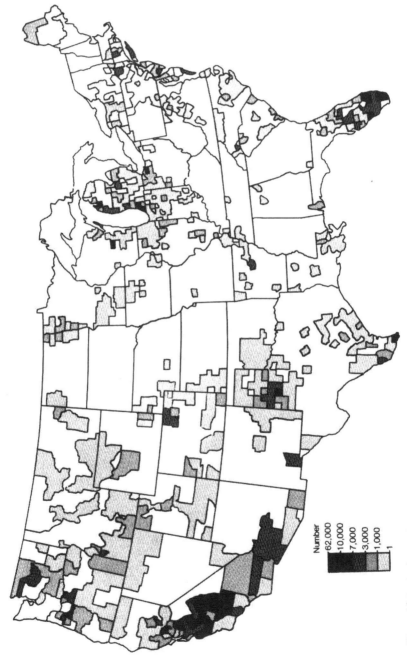

Figure 11.5. Agricultural migrants who migrated into counties, 1969.

Number
62,000
10,000
7,000
3,000
1,000
1

said Paul Hunter, executive director of Arizona's vegetable growers' association. According to Don McGaffee of Phoenix Vegetables, which cultivates 600 acres of green onions a year, "The atmosphere in the fields is not unwholesome. I can't see anything better than a kid being out in the open air. . . . To tell them they can't bring the children isn't right. The over-all income drops if the children stay home. This has been going on for years."[29] Arizona's labor department consists of ten employees, an inadequate number for enforcing child-labor laws. Seven professionals operate out of the Phoenix office and three out of Tucson enforcing age discrimination in employment and minimum wage, overtime, and equal pay as well as the child-labor provisions of the FLSA. No civil penalties—which can be as high as $11,000 for each violation—were levied against Arizona growers or contractors in the first two years of the new child-labor provision.

But inadequate staff to enforce the law is not really the heart of the problem. In 1978 the U.S. Department of Labor issued regulations allowing use of 10- and 11-year-old children in the harvesting of strawberries and potatoes, despite repeated warnings from government and private groups that the danger of pesticide exposure could result in lifelong damage to children— by causing cancer, damaging reproductive systems, and inhibiting growth. The regulation allows twenty-five pesticides to be sprayed on crops harvested by children.[30] Untested chemicals are being tested on children!

Although the exploitation of migrant children and the hazards they face are opposed by liberal politicians, the use of child labor on family farms is an accepted American tradition. The social relations within farm families obscure the issue of exploitation of child labor. Regardless of the proponents' arguments (outdoor work is healthy, work at an early age teaches life skills farm work promotes family togetherness) rural children are treated differently than urban children are. The use of child labor was once common in American cities too, but unionization and higher wages rendered this practice obsolete for industrial workers. Yet family farmers must continue to exploit their children because of the inadequacy of their farm incomes. U.S. agriculture must be restructured to allow family farmers to end this form of exploitation.

Working and Living Conditions

Farm workers are still denied the basic liberties granted to industrial workers in the 1930s by the National Labor Relations Act: to organize into unions and to bargain with growers over the terms and the conditions of their employment. The absence of collective bargaining has worked to the advantage of a few large-scale producers of certain crops in a few states. Most migrants are employed in some of the most prosperous agricultural regions and on some of the richest farms. Yet every aspect of the migrants' working and living conditions is inferior to that of industrial workers: wages, workers' compensation, unemployment insurance, fringe benefits, welfare benefits, social security, and housing.

Wages

The federal government has set a minimum wage for nonfarm work since 1938. But there was no minimum wage for farm work until 1967, and it has never equaled nonfarm rates. In 1976 the industrial rate was $2.30 per hour and the farm rate was $2.00. The rate for seasonal workers processing and packing agricultural produce was $2.30. By July 1979, the average hired farm worker received $3.23 per hour; field workers averaged $3.05 and livestock workers averaged $2.99. Because of the many exclusions in the law, only a fraction of all farm workers actually receive the federal minimum wage. The federal law excludes: (1) employees who work on farms with 500 or fewer person-days of hired agricultural labor during each quarter of the preceding year (this exempts most family farmers from paying minimum wages and allows them to extract even more surplus value from their hired hands); (2) workers who raise range livestock (this again exempts the smaller livestock operators in the West, as the largest ones are feedlot operations); and (3) piece-rate workers who commute daily to their permanent residences and who worked less than thirteen weeks on farms in the previous year. Growers using local people who occasionally work in agriculture therefore are subsidized by federal law exclusions and their workers are legally exploited more than other agricultural workers merely because of their residence and work experience, not on the basis of their productivity. Because of these exclusions, federal minimum wage rates extended to only 29 percent of the almost one million seasonal workers in the lower 48 states in mid-July 1971, and to only 38 percent of all 1.5 million hired farm workers.[31]

The Fair Labor Standards Acts specifically exempt agricultural employment from its overtime pay requirements. For growers these exemptions mean a great deal of profit, and for farm workers they mean exploitation. The arbitrary division between farm and industrial labor in legal wage scales and overtime pay reflects the class bias of the government. The needs of labor are forced to adjust to the needs of capital; the needs of the majority must accommodate the needs of the wealthy minority. The Jeffersonian ideal has been turned on its head.

The exploitation of farm workers is justified not only by legal exclusions but also by illegal evasions of the existing laws.[32] In 1970, Edgar Krueger testified before the Senate Subcommittee on Migratory Labor that in Texas, "piece-rate scales often fall far below the minimum wage level. Workers picking cotton now in South Texas seldom earn the minimum wage. When workers complain, employers threaten greater use of machines or say "if you don't like it, you can quit." . . . [T]he tally of hours worked [is made to look like] the minimum is being paid."[33]

For more than three decades the U.S. Department of Labor has operated and funded the Farm Labor Service, which cost almost a quarter of a billion dollars from 1962 to 1972. Its ostensible purpose is to assist farm workers in finding the best available jobs in accordance with federal health and safety standards. But, "in fact, its grower-controlled and -oriented staff provides the best workers at the lowest wages to the worst growers, those growers who, due to artificially low wage rates, are unable to compete in the open market-place for labor."[34] Consequently, wage rates are especially low for seasonal and migrant farm workers compared with other workers. The national average cash wage rate in 1976 for agricultural workers was $2.65 per hour, ranging from $2.20 in North Carolina to $3.10 in Washington. Workers who were paid piece rates averaged higher hourly earnings than workers who were paid by the hour. The piece rate averaged from $2.20 per hour in Louisiana to $3.35 per hour in California; the national piece-rate average was $2.94 per hour. But, contrary to popular belief, only a minority of seasonal farm workers were paid piece rates. In mid-May 1971, a time of cultivating and planting most crops, only 35 percent of all seasonal farm workers were receiving piece rates; by mid-July, the percentage had dropped to 25 percent.[35]

Apologists for agribusiness, such as Jerry W. Fielder, former secretary of agriculture for California, point with pride to the high wages migrants receive: "Wages paid to California farm workers have continued to increase and are among the highest in the Nation. The annual average composite hourly rate in 1970 was $1.87. This was 9 cents, or 5.1 percent, higher than the $1.78 per hour average for 1969. The 1970 California rate was also 45 cents, or 32 percent, above the national average of $1.42."[36] Although hourly earnings have increased in recent years, seasonal farm workers received only a fraction of the earnings of industrial workers. In July 1976 agricultural workers had an average hourly rate of $2.65, 27 percent above the average in July 1973.[37] This gain did not even keep up with the 20 percent inflation. The 1976 average cash wage of $2.81 per hour for migrants was only 57 percent of the average straight-time pay of production workers in manufacturing.

The increase in wages for seasonal agricultural workers over the years contrasts sharply with the increase in salaries of agribusiness executives and company profits. The California canning industry provides an excellent example of the exploitive wage system among agricultural processing firms. Heinz, Del Monte, Campbell Soup, and Libby, McNeill and Libby control 80 percent of California's $1 billion canning industry. The chief executives for the first three companies increased their salaries by nearly 100 percent from 1970 to 1975, while cannery workers were given pay increases of about 30 percent. The actual income difference between these two groups in 1975 was even more striking. On the one hand, the chief executives of Heinz, Del Monte, and Campbell Soup received $392,417, $226,066, and $553,798, respectively. On the other hand, the *highest*-paid cannery worker, who worked 60 hours a week for 28 weeks, earned $10,382, and the *lowest*-paid worker earned $6,535 (highest hourly rate: $6.18; lowest hourly rate: $3.89). In addition, the corporate profits of these three companies increased from 38 to 220 percent during the five years.[38]

Wage earnings of most migrants are exaggerated by even these low figures. Migrants are lucky to keep one-half or even one-third of their annual income. Crew leaders often control almost every facet of migrants' lives. They bring migrants north or east, find them jobs and housing, and often have the only trucks and cars. One migrant told his story of abuse:

He [the crew leader] charges us 75 cents for a pack of cigarettes (and we're the ones picking the tobacco for those cigarettes), and in this state [North Carolina] they only cost 35 cents in a store; he charges 25 cents for a cup of coffee—and what coffee!—and over a dollar for every drink, and it's no good wine. How can a man have any money left, with expenses like that? My first two weeks here I found out that I owed money to the crew leader. I work all week long, and who do I end up working for? Everyone but myself! Last week I finally got paid—$4.20. Four dollars and twenty cents. I came up here thinking I would make some money, and instead I'm going into debt faster than I can get out! And who's keeping the books! He [the crew leader] sits there and tells us what we owe, and that's it! That's freedom for you! Is there any difference now and when there was slavery?[39]

Farm workers are cheated by their employers as well. Testimony and case studies from Colorado, Florida, New Mexico, Washington, Minnesota, Iowa, and New York attest to the abusive power of growers, who withhold money illegally and pay at the end of the season after forcing migrants to use their company-issued coupons to buy groceries and supplies at high-priced local stores.[40]

In response to agribusiness, the seasonal farm labor force is changing in composition. According to the USDA's Economic Research Service, hired farm labor is increasingly young, includes fewer blacks, and works more on a part-time basis. Of nearly 2.8 million people hired to work on U.S. farms and ranches in 1976, 59 percent were under 25 years of age, an increase of 14 percent over the past 15 years. But although young workers made up the majority, most worked on a short-term, seasonal basis, performing only 45 percent of the person-days of hired work. In contrast, full-time farm workers did 65 percent of all hired labor, though they represented only 27 percent of the hired work force. In 1966 blacks made up 27 percent of all hired farm workers; by 1977 that figure had dropped to 17 percent (see Figure 11.6). Altogether, minorities accounted for a fourth of all persons hired for farm work, but made up 38 percent of the workers primarily dependent on farming for their livelihood.

Farm work has also become a secondary occupation for many migrants. In 1976, nonfarm work was the primary source of income for 17 percent of those hired, up from 12 percent in 1966. They earned an average salary of about $6,000 from farm sources. Those who held jobs in both areas but relied mostly on farm work earned roughly $4,800 Working exclusively on farms meant an average yearly income of about $5,200 (see Table 11.4).[41]

Agribusiness manipulates seasonal workers to its advantage. First, growers fix wages among themselves. Lloyd Fisher, a labor economist, stated that "wage fixing by employers is taken as a matter of course in California and the terms of these agreements are disseminated by the public press or by association bulletins or by the journals of farm organizations and commodity associations."[42] Second, growers, in alliance with state and federal agencies, keep wages down by recruiting more labor than they need. Responding to

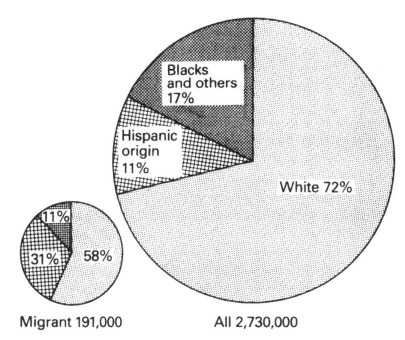

Figure 11.6. Racial/ethnic background of hired farm workers, 1977.

TABLE 11.4
Average Annual Earnings of U.S. Farm Workers, 1976

Major Employment Status	Total Number	Percent of Farm Work Force	Average Annual Earnings($)		
			All Workers	Persons with Farm Work Only	Persons with Nonfarm Work[a]
Primarily Farm Workers[b]	746,000	27	4,762	5,176	4,727
Primarily Nonfarm Workers[c]	474,000	17	6,030	---	6,030
Students	1,089,000	39	926	644	1,375
Homemakers	244,000	9	817	588	1,365
Others[d]	214,000	8	1,368	607	2,618
Total	2,767,000	100			

[a]Combined farm and nonfarm earnings.
[b]People who are in the labor force most of the year and employed as hired farm workers more than half the time.
[c]People who are in the labor force most of the year and employed as hired farm workers less than half the time.
[d]Includes people who were unemployed most of the year.

Source: USDA, Agricultural Situation, August 1977, p. 8.

advertisements of a "shortage of labor," foreign workers, school children, and homemakers are attracted into agriculture, thus reducing their own wages and the wages of other agricultural workers. Agricultural workers who make a living following the crops would more accurately label this situation a "shortage of pay."

Contrary to popular belief, growers would still profit if farm workers received higher wages. Stephen Sosnick, an agricultural economist, showed that "for the United States as a whole, growers' combined net income would have remained positive even if every grower's cost of hired labor had tripled."[43] The corollary myth that higher wages would force consumers to pay higher prices for food is also unsupported (see Box 11.1). The cost of field labor is only a minute part of the retail price paid by consumers. On a head of lettuce selling for 21 cents in 1965, the field labor cost was 1 to 1.3 cents. On a pound of celery retailing at 15.5 cents per pound, the cost of field labor was 0.3 to 0.5 cents. On oranges retailing at 50 to 72 cents per dozen, the field labor costs were 1 to 2 cents.[44] Tripling farm-worker wages would hardly be felt by consumers, who pay much more for other costs of agribusiness.

Workers' Compensation

Farm workers face not only the uncertainty of employment—seasonal at best—and low pay when they do find work, but also the hazardous working conditions of agriculture without the protection that other workers receive. Workers' compensation, the oldest form of social insurance, provides payments to workers who are injured on the job. These payments include death benefits, medical costs, and a weekly pay check amounting to about 70 percent of workers' predisability wages for as long as they are disabled. But workers' compensation, *required* for most American workers, is not provided for farm workers. The need for such protection is particularly compelling in agriculture, which has the third highest disability injury rate, ex-

---

BOX 11.1. THE EXPLOITATION OF JAMAICAN
APPLE PICKERS IN VIRGINIA

"The prime Red Delicious apples that pickers harvest for two-thirds of a penny per pound bring growers about twelve cents a pound, the packers about seven cents, the transporters a penny or so, the wholesalers two or three cents, the brokers a penny or two. The retailers add a markup of at least 100 percent. The consumer in mid-winter thus pays fifty cents for two or three apples. That translates to about $22.50 for a forty-five-pound box of fruit that a picker harvested a few months earlier for thirty cents."

Source: John Egerton, "As Jamaican as Apple Pie," *The Progressive*, December 1977, p. 40.

ceeded only by construction and mining. Agriculture also has the highest occupational disease rate, caused mostly by fertilizers and pesticides. Furthermore, farm workers face particularly severe working conditions. They are often sent into freshly sprayed fields and do some of the most dangerous and laborious work in the fields and in processing plants. Racist and classist attitudes help employers justify such treatment of workers, and the lack of workers' compensation provides legal validation of these attitudes and practices at no expense to growers.

Unemployment Insurance

About five months out of each year, from November to April, seasonal farm workers cannot find work and must compete for the few available jobs in their home counties of southern Texas, Florida, and California. During the 1960s only 35 percent of wage and salary workers in agriculture were able to work a full year, compared with 59 percent of workers in nonagricultural industries. The migrants who were able to find nonfarm work—perhaps two-fifths of them—averaged only 8.5 months of paid employment, while the average migrant who did only farm work reported about 7 months of paid work.[45]

By 1974, 87 percent of all *nonfarm* wage earners in the United States were covered by unemployment insurance. Because employers contribute to this insurance, few voluntarily participate in the program. In 1968, eight years before California extended mandatory coverage to farm workers, less than 2 percent of the 50,000 farms in the state made voluntary payments, and these farms employed less than 10 percent of the state's 200,000 hired farm workers.[46] Seasonal farm workers themselves were usually against this program because they had to contribute to it, which reduced their low earnings even more.

In 1978, by federal law, wages paid for farm work became subject to unemployment insurance taxes, but only if employers had payrolls of at least $20,000 or had ten or more employees during a twenty-week period. Although states may impose unemployment insurance taxes on smaller farm employers, only three states—Hawaii, California, and Minnesota—have done so.[47]

Fringe Benefits

Farm workers do not receive the fringe benefits that are routine for most other workers, particularly unionized workers. Over 73 percent of all hired farm workers employed on large farms (hiring seven or more full-time workers) received no benefits in 1971. If smaller farms are included, the proportion of workers receiving no benefits is much higher.[48] In contrast, about 80 percent of industrial workers have company pension programs; paid vacations and holidays; and company-financed life, hospital, and medical insurance. Although nonfarm companies pay 26 percent of their cash wages in wage supplements not required by law, growers pay only 6 percent of their cash wages for such supplements.[49] Since farm labor is more productive than industrial labor, these differential fringe benefits can serve as one indicator of the degree of farm-labor exploitation—in this case, roughly four times that of nonfarm labor. The absence of fringe benefits, particularly health benefits, is reflected in the health of migrants. Average life expectancy is 49 years, compared to 70 years for other workers. The infant mortality rate for migrant families is 125 percent higher than the national average, the maternal mortality rate is 118 percent higher, and tuberculosis and infectious diseases are 260 percent higher.

Welfare Benefits

Although most farm workers must contend with seasonal unemployment, low wages, and the absence of benefits, they seldom rely on welfare. A Colorado study found that only 8 percent of Mexican-American/Chicano migrant families had received any form of financial assistance from welfare departments and none had received food stamps.[50] These people do without, not because of pride or lack of need, but because legal restrictions have prevented seasonal farm workers, particularly migrants, from obtaining welfare benefits. Aid to families with dependent children (AFDC) requires a permanent dwelling place—and this requirement automatically disqualifies all 650,000 migrants and their dependents. AFDC is also denied to families with employable fathers, whether or not they are employed. Because of this eligibility requirement, low-income fathers are forced to nominally abandon their families so that the mothers can receive welfare payments for their children. Thus the law makes criminals out of honest people! Farm workers who overcome these two obstacles face more delays in getting aid. The application procedure is particularly difficult on seasonal farm-worker families with impermanent homes and low wages. But the welfare laws are written by middle- and upper-class politicians and administered by middle-class bureaucrats, whose interests do not accommodate the needs of those for whom these laws were supposedly passed.

General assistance, another form of welfare, is rarely received by migratory farm workers. Because federal emergency relief was given directly to strikers during the middle 1930s, workers had the ability to hold out for higher wages. Thereafter, largely due to pressure from agricultural interests, welfare payments were brought under state and county supervision. At the state and local levels, growers knew they could control the availability and conditions of general assistance. Many counties with large numbers of interstate migrants— all Texas counties, for example—have no general assistance at all. In Florida, another important home base for migrant families, small amounts of general assistance are available in some counties and none in others. Inadequate funding, residence requirements, and employment criteria are used to exclude the most needy of farm workers—migrant families.

The food stamp program is another source of assistance for America's poor. In the 1960s few migrants received food stamps. Most help was received in their home states; less help was forthcoming on the road. One report concluded:

> Less than 16% of the migrants in the [Texas] counties studied were served with food assistance programs [in 1969]. Yet migrants fared

better in Texas than in any of the stream states [Colorado, Illinois, Indiana, Michigan, Ohio, Oregon, Washington, and Wisconsin]. In Michigan, for example, less than 2% of the migrants in any county studied were included in food programs; in Wisconsin less than .001% were included.[51]

By 1974 all 3,129 counties in the United States had finally participated in the food stamp program. Yet the USDA estimated in mid-1975 that 29.2 million people were eligible for food stamps in any one month, yet only 59 to 67 percent (17.2 to 19.5 million) of these people received stamps[52] (see Figure 11.7). Although no separate figures were given for migrants, the conditions under which they live and the restrictions of the programs would undoubtedly result in much lower participation rates for them.

Social Security

As a marginal class, farm workers gained social security later, faced greater legal restrictions, and suffered more abuses from employers than industrial workers who gained coverage in 1936. Year-round and seasonal farm workers were covered in 1950 and 1956, respectively. Workers outside agriculture are

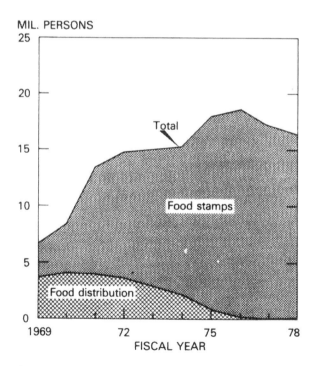

Figure 11.7. Participants in the family food assistance programs, 1969–1978.

subject to social security taxation when they earn more than $50 in any calendar quarter, but farm workers must earn at least $150 in cash wages from *one* employer during a calendar year or work at least twenty days for cash, regardless of the amount earned. The latter restrictions penalize students and homemakers who tend to work in agriculture for only brief periods. Moreover, the $150 or twenty-day threshold encourages growers and processing-plant employers to dismiss workers before the threshold is reached. In this manner, growers avoid their matching contribution and pocket millions of dollars that should benefit agricultural workers. In addition, workers who quit, disappear, or are discharged frequently lose their social security contributions. Seasonal farm workers, especially those who work by the day, receive no records of earnings or receipts for deductions and therefore have no protection against unscrupulous farmers or crew leaders who may illegally appropriate their social security contributions. Thus deprived of all retirement income, migrants are condemned, after a lifetime of hard labor, to old-age poverty even more severe than that of the millions of Americans who barely survive on their social security checks.

Housing

Housing for migrants, either in their home counties or on the road, rarely meets minimum standards of health, safety, and sanitation. Sixty percent (4.8 million units) of all substandard housing in the United States is located in the countryside, yet less than 25 percent of all federal housing funds have gone to rural housing.[53] In 1971, 18 percent of migrants lived without electricity, 90 percent without sinks, and 96 percent without baths. Still, the wife of a local police chief thinks the living conditions of black migrants in the Northeast are adequate: "This place is a paradise compared to what they are used to living in. Of course you or I wouldn't want to live that way, but I believe they like it fine."[54]

For the most part, migrants and their dependents live in dilapidated, drafty, ramshackle structures that are cold and wet in winter and hot in summer. In many cases, substandard housing is provided by growers. In Oregon and Washington, 75 to 85 percent of the housing units were located on fruit and vegetable farms where migrants worked.[55] In other cases, migrants must find off-farm housing in labor camps operated by labor contractors or by the federal government or in adjacent towns at commercial rates and subject to economic, racial, and occupational discrimination.

In the summer of 1976, students from Duke University examined migrant labor camps in North Carolina. Their findings are indicative of similar conditions in other states. Over 400 migrant labor camps accommodated approximately 10,000 migrant workers who made their way into North Carolina during the first part of June, when cucumbers were ripening, and stayed until late November, when the last of the sweet potatoes were harvested. In almost every camp health and safety standards were not met; one Occupational Safety and Health Administration (OSHA) official acknowledged that 95 percent of the known camps were in violation of the law. Of the 400 known labor camps, only 30 percent were inspected by OSHA and only 15 percent

Figure 11.8. Migrant housing for black agricultural workers in Pahokee, Florida, 1941. Marion Wolcott, FSA photo.

Figure 11.9. A labor camp in California's Central Valley: Yuba City, 1976.

of those camps were inspected a second or third time at a later date (an absolute necessity if standards are to be enforced). For every officially registered camp there was another not inspected. In total, only 1.5 to 2.25 percent of the total number of labor camps in North Carolina were given more than a single, cursory inspection. Often government officials know little about their clients and seldom visit the camps in which the programs operate, working instead through a few subordinate field instructors.[56] Furthermore, OSHA uses industrial safety standards rather than housing standards to judge migrant housing. If the U.S. Department of Labor continues to regard migrant housing as a network of factories, then labor camps will, at best, become "injury proof" tenements. These standards can hardly be sufficient for places of human habitation.[57]

The Inadequacy of Liberal Programs

Over the years, when the liberal conscience of the American public and Congress has been stirred by the plight of farm workers, political action has often resulted. In 1962, for example, the Migrant Health Act was passed with an appropriation ceiling of $3 million for a three-year period and was subsequently extended several times. Before its passage, members of the Senate Subcommittee on Migratory Labor made personal observations during extensive field visits to migrant areas, and their hearings confirmed over sixty years of reports—that migrant farm workers were "a forgotten minority, the neediest and the least served."[58] Similar pieces of legislation providing services in nutrition, education, job training, day care, housing, and legal aid have also been passed.[59] These liberal federal programs make the effort to empathize with the wretched of the earth but fail to identify the root causes of wretchedness in agribusiness. Ernesto Galarza stated the inadequacy nicely:

> Some of the power of the liberal conscience lies in its idealization of the miserably exploited, for whom it arouses compassion. Its weakness lies in its inability or disinterest in helping class victims create their own collective means for resisting abuse and for changing the conditions that cause it. The liberal conscience drops the task when only half completed, its protégés still incapable of taking their rightful and responsible place among the free.[60]

A 1973 General Accounting Office report declared that the programs of the Department of Labor; Department of Agriculture; Department of Health, Education, and Welfare; and Office of Economic Opportunity had had little appreciable impact on the lives of millions of farm workers. The report noted that although the government had spent in excess of $650 million in grants and loans to individuals and organizations working with seasonal and migrant farm workers, farm workers for the most part remained ill-housed, poorly educated, and untrained and received inadequate medical treatment.[61] Since 1962, when the Farm Labor Housing Loan and Grant Program was first enacted, the Farmers Home Administration has committed $31 million to finance housing for 4,700 families and 3,456 individuals. Over a ten-year period, this amounts to a yearly average of 470 families and 345 individuals for whom housing has been improved, out of a total migrant and seasonal farm-worker population numbering in the millions.[62] This and other migrant-related programs failed in large part because the programs, even if federally funded, are locally administered. Local political patronage determines the way migrants are treated, and migrants have no power in this patronage. For example, in Johnston County, North Carolina, the health inspector is the son of a grower and the sheriff and a county commissioner are brothers of growers.[63]

Migrant Response to Agribusiness:
The United Farm Workers of America

In the context of California's heavy reliance on Mexican-American/Chicano labor, racism and traditional class antagonism have fostered the most persistent attempts to organize and unionize farm workers in U.S. history.[64] The geography of California has facilitated this class struggle: mild weather and fertile valleys mean that almost anything can be grown profitably. Agriculture is year-round, with much of the land yielding two, three, and or even four crops per year. Even with the introduction of labor-saving machines, new land is being "created" through irrigation (there were eight million irrigated acres in 1972, twice as many as in 1939). Farm labor has also increased to cultivate this expanded productive base. In 1974, the state had a farm employment of 287,200, an increase of 7,500 over 1972.[65] (See Figures 11.10 and 11.11 for the extent of cropland and employment potential in California.)

The historical concentration of farmland among a few landowners in California has required large amounts of hired labor (more than 15 percent of California's farms have wage labor payrolls larger than $20,000 a year).[66] United Brands, a Boston-based conglomerate, owns the largest lettuce farm in the country (Interharvest) and cultivates a total of 22,000 acres in California and Arizona. Tenneco farms 1.4 million acres in these two states. One third of all the wine consumed in the United States is produced by Ernest and Julio Gallo, who own 3,500 acres of vineyards. Even those farms that can be called family farms extend for thousands of acres. The conditions for a protracted labor struggle in agriculture have had plenty of opportunity to develop, and indeed it is here that farm workers have continually challenged agribusiness.

Attempts to organize farm workers by the Wobblies at the turn of the century, by Communist-led unions in the 1930s, and by AFL-CIO–backed committees in the 1950s were short-lived. All too often organizers were outsiders with little understanding of farm workers; local, state, and federal authorities collaborated with growers; racial mistrust among workers kept them divided; strikebreakers were too numerous; and strike funds were inadequate.[67]

The most recent attempt to organize farm workers has been the Farm Workers Association (FWA), which had several thousand dues-paying members in 1965. When Filipino farm workers called a strike against table grape growers in the Delano area, the predominantly Chicano FWA unanimously voted to

Figure 11.10. Crop areas (shaded) in central California using migratory labor. *Source:* U.S., Department of Labor.

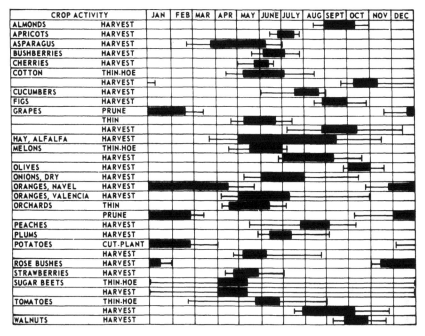

Figure 11.11. Employment potentials for seasonal farm workers in the San Joaquin Valley, California. *Source: California Guide for Farm Workers*, California Rural Manpower Services, Department of Human Resources Development, n.d.

join them. In 1970, after five years of strikes and boycotts, the renamed United Farm Workers of America (UFWA) won its first major contracts with most of the major wineries, the lettuce-growing industry of United Brands, 85 percent of the table grape growers, a significant number of strawberry growers, and the citrus-growing subsidiary of Coca-Cola, Minute Maid. These contracts not only substantially raised wages of field workers, but also improved the working and living conditions of farm workers. UFWA-run hiring halls replaced the notorious contracting system operated for the benefit of labor contractors and growers.[68] Sanitary facilities in the fields were improved, and strict rules regarding the use of pesticides were established and enforced by the UFWA. The union also opened five medical clinics and built a retirement village for Filipino farm workers.[69]

The UFWA was putting into practice its goals to reform the basic nature

DON'T SWALLOW GALLO'S LINE!

You may have seen ads, pamphlets or letters recently from the E&J Gallo wine company, talking about the farm labor situation. Rather than confront the truth and justice of the farmworkers' charges against them, the brothers Gallo have launched a massive PR offensive aimed at obscuring the issues and misleading the public.

The facts of the Gallo situation speak for themselves:

In 1967, Gallo signed with the United Farm Workers (UFW) on the basis of signed authorization cards from a majority of the workers, verified by the California Department of Industrial Relations' Conciliation Service.

In 1973, Gallo signed with the Teamsters, on the basis of Petitions "verified" by no one except Gallo management. Even as Gallo was saying its workers wanted to change unions, a delegation of priests and nuns

this was our housing at Gallo. They finally tore it down.

was offering to present signed UFW authorization cards from a majority of the workers. Gallo ignored their phone calls and telegrams.

Cesar Chavez sent Gallo a telegram requesting free elections to settle the dispute. Gallo and the Teamsters refused to allow elections, for obvious reasons.

The Teamster contracts were substantially inferior to UFW contracts at three smaller wineries — Almaden, Christian Brothers and Novitiate. Guarantees on sanitation, pesticide protection, rest breaks, and prohibition of child labor were either eliminated from the contract or were rendered meaningless by the lack of worker-supervised enforcement.

this was our communal bathroom. Gallo couldn't afford stalls.

Gallo claims its full-time workers average $7,785 a year, and seasonal workers average $278 per week. They can undoubtedly produce a few check stubs showing apparently high wages, but only because it is the practice to issue a single check for the work of an entire family. The UFW has in its possession a check stub from a Gallo worker who worked 27 hours and, after various deductions—including $56.98 for "miscellaneous"—took home a grand total of $1.10. Even if you accept Gallo's claims at face value, their $2.89 per hour minimum wage is lower than minimum wages at Almaden, Christian Brothers and Novitiate.

When Gallo ignored its workers' wishes and signed with the Teamsters, most of the workers went out on strike. Gallo fired them, replacing them with strikebreakers and illegal aliens. The new work force, not surprisingly, voted to accept the new contract rather than lose their jobs. But the original work force, on the payroll at the time the previous contract expired, was never given the chance to vote on which union they wanted, let alone to ratify the new contract.

In any other industry, Gallo's tactics — changing unions without consulting the workers, bringing in new workers to negate a strike - would be illegal. But agricultural workers aren't covered by the protections all other American workers enjoy. Which leaves them just about powerless — unless you help.

By refusing to buy any Gallo wines*, you can help the workers win the right to belong to the union of their choice. Buying some other wine won't make much difference in your life — but it'll make a big difference in the lives of thousands of farmworkers.

... after all, Cadillac prices have gone up. this is where the Gallo brothers work.

* Boone's Farm, Madria Madria, Tyrolia, Ripple, Thunderbird, Spanada, Wolfe & Sons, Andre, Paisano, Carlo Rossi, Red Mountain, Eden Roc and *any* wine made in Modesto, Ca. are Gallo.

DON'T BUY THE GALLO LINE

BOYCOTT ALL GALLO WINES!

UNITED FARM WORKERS OF AMERICA, AFL-CIO

Figure 11.12. The United Farm Workers fight against agribusiness. Reprinted by permission of the United Farm Workers of America, AFL-CIO.

of large-scale agriculture: protect members against pesticides and mechanization; build a political force for Chicano and Filipino farm workers that could offset the political power of agribusiness; and nurture a sense of solidarity, self-respect, and well-being among the poorest and most invisible farm workers, nationals and foreigners alike.[70]

In 1973, the majority of growers, who had signed with the UFWA, switched to the Teamsters. Since farm workers were and still are not covered by the National Labor Relations Act, growers could switch contracts without consulting their workers. This grower tactic was successful because it weakened the UFWA, both in membership and finances (by 1974 membership dropped from about 60,000 to 12,000).[71] The few resources of the UFWA had to be stretched to cover four fronts: growers had to be won over again; farm workers, most of whom were transient and desperate for jobs, had to be reached again; the food-buying public had to be convinced to participate in a national lettuce, grape, and wine boycott to force the growers to negotiate with the UFWA (see Figure 11.12); and the two-million-member Teamsters union had to be challenged. When contracts by the UFWA and the Teamsters were compared, it became clear why growers switched to the Teamsters without the consent of their workers (see Table 11.5).

While the UFWA's national boycott of lettuce, grapes, and wines was having its full effect, California growers and major farm associations supported state legislation that would give farm workers the right to choose their own union. They were conceding not to the farm workers themselves but to effective mobilization of U.S. consumers by the UFWA. At the height of the boycott, a Louis Harris poll reported that 12 percent of those questioned were boycotting table grapes; 11 percent were boycotting lettuce; and 8 percent had stopped buying Gallo wines. Projected across the national population, this meant that 17 million people were part of the grape boycott, 14 million joined the lettuce boycott, and 8 million supported the more recent effort against Gallo.[72]

The UFWA-initiated boycott was effective not only because large numbers of people participated but also because they drove profits down. Growers admitted getting only $1.70 for a $2.50 box of lettuce, an 80-cents loss per box! Grape growers lost between $15 and $20 million. Gallo's national sales went down 20 percent while UFWA wines went up as much as 250 percent. Gallo spent $13 million on television and magazine advertisements countering the boycott[73] (see Figure 11.2).

In the summer of 1975, Governor Jerry Brown signed into law the Agricultural Labor Relations Act. By the end of that year's harvest season, the UFWA had won 176 elections accounting for 25,339 laborers, and the Teamsters had won 96 elections representing 12,802 farm workers. At 18 ranches, employing 11,505 workers, no union had been selected.[74] By February 1976, the UFWA was the clear choice on 193 ranches and the sole bargaining agent on another 80 ranches.[75] The union had won over two-thirds of all elections held in California fields under the state farm-labor law by the spring of 1977.[76]

TABLE 11.5
Comparison of Contracts Between the UFWA and Teamsters in 1974

Contract Item	United Farm Workers of America	International Brotherhood of Teamsters (with Gallo)
Contract Covers	All employees working on a company's owned, leased, or rented fields	Only employees working directly for company; employees supervised by labor contractors would be excluded--about 90 percent of farm workers.
Labor Contractors	Not permitted	Permitted
Hiring	Union-run hiring halls direct workers to company on the basis of seniority	Labor contractors select farm workers for companies. Employers responsible for seniority, but not required to reveal lists.
Probationary Period	None	First 30 days; if fired during this time, workers have "no recourse to the grievance procedure."
Seniority	Days worked on a particular farm; seniority cannot be lost between seasons.	Last hired, first fired. Seniority can be broken if workers are laid off "for 6 months or for a period of time equal to their seniority, whichever is shorter."
Discrimination	None allowed, including political beliefs	None, but political beliefs
Hourly Wage Rate	15 classified wage rates: lowest $2.95; highest, $3.85 per hour	10 classified wage rates: lowest, $2.89; highest, $3.75 per hour. More workers receive the lowest wage rate.
Pay on Sundays	Time and a half	Regular hourly rate plus 25 cents per hour
Eligibility for Holidays	All employees who worked 1,000 hours in the prior calendar year	Only regular full-time employees who worked 1,500 hours in the prior calendar year
Days off Without Pay	One full day without pay each payroll week (six-day work week)	"No limitation on the daily or weekly hours that employees are required by the company to work."
Leave of Absence for Funerals	All workers receive three days at the regular hourly rate for deaths in the immediate family.	Only regular full-time employees receive three days of leave for funerals in their immediate families. This covers only 10 percent of the workers.
Rest periods	15 minutes in the morning and in the afternoon	10 minutes for each four hours of work, "insofar as practical"
Introduction of New Machines	No, as long as workers are available	Yes
Maintenance of Standards	"Wages, hours of work and general working conditions shall be maintained at no less than the highest standards."	"The sum of wages and benefits (excluding housing) that employees are presently receiving...shall not be reduced."
Company Housing	Workers pay no rent	Workers who started after September 21, 1967 pay a minimum of $95 per month. This excludes 92 percent of Gallo workers.
Protective Clothing Against Pesticides	Company provided	No provision
Toilets in the Fields	Company provided	No provision
Drinking Water in the Fields	Company provided	No provision
First Aid Supplies in the Fields	Company provided	No provision
Medical Plan	Workers are covered no matter how long they work on any given ranch under contract. Family members can pool their hours for benefits.	Workers are only covered if they worked 80 hours for the company in the preceeding month. Family members can not combine their hours for eligibility.

Source: The Facts: Why Workers Want the UFW-AFL-CIO and Why Growers Want the Teamsters. (Keene, California: United Farm Workers of America, 1974).

12
The Decline of Agriculturally Dependent Small Towns

Agricultural researchers have shown little interest in the social consequences of large-scale farming, which they helped create, despite massive research efforts to study the relationship between farm size and efficiency. By researching aspects of productivity without regard to human consequences, these scholars reflect and perpetuate the ideological bias of agribusiness and the monopoly capitalism from which it springs.

The absence of studies examining the effects of large-scale farming on the prosperity and vitality of small U.S. towns is striking, especially in view of the negative impacts of agribusiness on local communities. Large farms, particularly company-run farms, generally buy farm equipment and production supplies at discounted prices directly from either wholesalers or factories, bypassing local retail establishments. For example, in 1977 Shinrone, which operates a large farm in Sac County, Iowa, purchased $250,000 in farm equipment directly from companies in Brantford, Canada and Detroit, Michigan and Algowa, Wisconsin. When farms are owned or controlled by vertically integrated companies, such as Tenneco or Purina, these farms buy directly from the parent company. This practice is doubly profitable for the companies and doubly destructive for the society as a whole: corporations strengthen their monopolistic control over the national economy and local businesses are bypassed. Even when large-scale farmers purchase supplies locally, they often manipulate local competing dealers to receive the lowest prices—at little or no profit to the merchants. Over the years, these practices destroy the economic viability of small agricultural service centers.

Large-scale farmers more often obtain credit, insurance, and legal services in distant cities rather than buying these services locally, as smaller producers are likely to do. A rural banker outlined the problem:

> The rural community lives from the gross income of the family farm or the small, closely held family farm corporation. Because towns and banks are in the business of serving people, the banker sees that the disappearance of these families would cause his town and his bank to disappear. . . . The fact remains that the small town can not exist without people on the land, no matter how productive a vast corporation farm may be.[1]

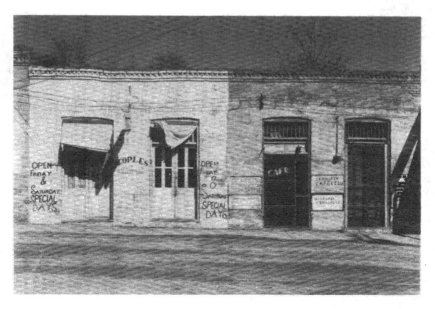

Figure 12.1. Agribusiness brings the demise of agriculturally dependent towns: store fronts on Main Street, Edwards, Mississippi, 1936. Walker Evans, FSA photo.

In contrast, where family farms thrive, small businesses flourish, too. A 1967 study by the Farmers Home Administration found that 190,000 farm families, who had received credit from the agency, grossed $3.2 billion, all of which was spent locally: $736 million for clothing, food, and other consumer items; $1.7 billion for goods and services to produce crops and livestock; and $704 million to retire debts and buy new farm machinery. When farm customers decline, small businesses also decline. One small business, on the average, is forced to close its doors every time six farm families leave the trading area.[2]

Few studies show the consequences of farm size on the quality of small-town life. Walter Goldschmidt's 1944 study of Arvin, a large-scale farming area, and Dinuba, a small-scale family farming area, remains the only comprehensive analysis of its kind.[3]

Arvin and Dinuba in the 1940s

Goldschmidt, trained as an anthropologist, selected Arvin and Dinuba because they had essentially the same geographical base and agricultural production but differed in the size of landholdings and the scale of production. Both towns are located in the San Joaquin Valley of the Central Valley in California. Each has the same locational advantages: Arvin is fifteen miles southeast of Bakersfield and Dinuba is thirty miles from Fresno. Both have about the same soil conditions and growing season and get only about ten inches of rain per year; thus, both rely on irrigation. Both agricultural communities had a diver-

TABLE 12.1
Arvin and Dinuba Compared: Farming Characteristics, 1944

Characteristic	Large-Scale Farms Arvin		Small-Scale Farms Dinuba	
Cultivated Area	22,000 acres		24,000 acres	
Gross Farm Sales (1940)	$2,438,000		$2,540,000	
Average Farm Size	497 acres		57 acres	
Average Value of Production	$18,000		$3,400	
	Number of Farms	Total Acres	Number of Farms	Total Acres
Average Size				
Under 80 acres	47%	9%	89%	56%
80-640 acres	47	51	10	35
Over 640 acres	6	40	0.1	9
	100	100	100	100
Farm Tenure				
Full Owners	35%		75%	
Part Owners	23		8	
Tenants	42		14	
	100		100	
	Million Person-Hours			
Total Labor Requirements	2.9		3.5	
Hired Farm Labor	2.3		1.4	
Wage Labor/Household Heads	81%		49%	
	Person-Hours in			
Peak Season Labor Demand	July		September	
Farm Operators	6%		27%	
Resident Laborers	36		16	
Part-time Resident Laborers	13		9	
Migratory Farm Workers	45		48	
	100		100	

Source: Walter R. Goldschmidt, "Small Business and the Community:
A Study in Central Valley of California on Effects of Scale of
Farming Operations," reprint ed in U.S., Congress, Senate, Role
of Giant Corporations, Hearings before the Subcommittee on Monopoly of the Select Committee on Small Business, 93d Cong., 1st
sess., 1973, part 3A.

sity of crops in the 1940s: Arvin emphasized row crops, like cotton, and Dinuba specialized more in fruits and vegetables. The total cultivated areas were essentially the same, as were gross farm sales.[4] Although both communities had gross farm sales of about $2.5 million in 1944, the distribution of this wealth and its consequences on local town life were fundamentally different (see Table 12.1).

Farming Characteristics

Six percent of Arvin's farms were larger than 640 acres and held 40 percent of the total acreage; 89 percent of Dinuba's farms were under 80 acres and held 56 percent of all land. Farm tenure and farmland ownership also varied dramatically: in Arvin 42 percent of the farmers were tenants, in Dinuba only 14 percent. Full owners who lived on their farms were the predominant pattern in Dinuba, while part owners, tenants, and absentee ownership were more common in Arvin. Consequently, wage labor was more important in Arvin than Dinuba.

Socioeconomic Characteristics

The socioeconomic characteristics of people living in Arvin and Dinuba reveal negative consequences of large-scale farming and positive consequences of small-scale farming (see Table 12.2). Although Arvin had a larger in-town population than Dinuba, only 9 percent of Arvin's employment consisted of white-collar workers. Since over half of Dinuba's employment base consisted of people who had access to resources—land for farmers and formal education for white-collar workers—incomes were higher in Dinuba than Arvin. Indeed, even laborers did better in Dinuba than in Arvin; 40 percent of them had incomes above the Dinuba median, but only 36 percent of the Arvin laborers reached their town's median income.

The class structure of these two communities reflects the distribution of landownership (see Figure 12.2). In power-concentrated communities, wealth is concentrated among the farm owners and professional classes, and over three-quarters of the population consists of relatively poorly paid laborers. Farm laborers are even less powerful than other workers. In power-diffused communities, on the other hand, wealth is more widely shared; farm operators are three times as common and white-collar workers are twice as important. Although laborers consist of almost half the population, more of them work in nonfarm occupations than on farms. Evidence from many other rural studies confirms this conceptualization of the class structure in large-scale and small-scale farming communities.[5]

Town Characteristics

The difference in the quality of life in these two communities is striking (see Table 12.3). Arvin had fewer elementary schools and experienced higher teacher turnover rates than Dinuba. The number of trade and social organizations and participation in these groups was lower in Arvin than Dinuba. The same was true of religious services. Dinuba had more civic pride and social participation in the affairs of the town because it had more family farmers who owned and lived on the land they worked. Arvin, with its higher percentage of laborers and absentee owners, showed less community participation. Goldschmidt found that laborers did not participate in civic clubs because social barriers existed for them nearly equally in both communities.

On virtually identical resource bases, measured by dollar value of gross farm sales, Dinuba merchants did approximately $4.4 million worth of retail trade compared to about $2.5 million among Arvin merchants. Dinuba had more than twice as many businesses as Arvin and offered different kinds of retail shops. There were more eating places, clothing and luxury shops, and home furnishing stores in Dinuba, and almost twice as many agricultural equipment and supply stores. Still, Arvin had a larger proportion of businesses

TABLE 12.2
Arvin and Dinuba Compared: Socioeconomic Characteristics, 1944

Characteristic	Large-Scale Farms Arvin	Small-Scale Farms Dinuba
Population		
Town	4,042	3,790
Open Countryside	631	3,877
	4,673	7,667
Occupational Structure		
Farm operators	11% ⎫ > 19%	34 ⎫ > 51%
White-collar workers	9	17
Nonfarm laborers	16	20
Farm laborers	65	29
	100	100
Family Head Income		
Less than $2,800	79%	65%
Greater than $5,300	8%	11%
Median income	$2,100	$2,350
Income Above Median		
Farmers	79%	65%
White-collar workers		
Nonfarm laborers	36%	40%
Farm laborers		
Educational Attainment		
8 years or less	77%	60%
9 to 12 years	17%	30
13 or more years	6	11
	100	100

Source: See Table 12.1.

with high sales volume and lower capital investment than Dinuba. The returns by merchants on the basis of their investments were therefore higher in Arvin than in Dinuba, but the lower financial investment by Arvin merchants was reflected by their lower social investment in community affairs.

Obviously, the merchants of Arvin were responding to the occupational structure of the community. Large-scale farms, with absentee ownership and heavy use of farm workers, create towns with fewer economic and social services than towns surrounded by moderate-sized farms. When merchants are in the business of extracting as much surplus value (profit) as quickly as possible, the social consequences for the quality of life in the town are devastating. Deploring such consequences is based on the assumption that

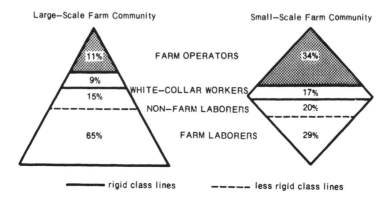

Figure 12.2. Class structure of communities dominated by agribusiness and family farmers.

physical comfort, material possessions, economic security and well-being, and social democracy are desirable qualities in any rural community. Even proponents of large-scale farming and agribusiness knew that these were the values Americans espoused and were trying to achieve, yet Goldschmidt's findings were so threatening to agribusiness interests that they tried to suppress his report. Agribusiness proponents in Congress put pressure on Goldschmidt's superiors to suppress the first part of the study when it was completed and to prevent the second part, which was to examine the entire San Joaquin Valley, from being undertaken. Without the efforts of Senator James A. Murray of Montana, who personally got the suppressed report released by promising not to attribute it to the USDA, Goldschmidt's study would not have been made public! But the hegemony of agribusiness succeeded in forcing Goldschmidt out of government service and, several years later, in getting Congress to abolish the sponsoring agency, the Bureau of Economics. The mission of that agency, to examine the social consequences of industrial farming, has never been resumed by the USDA. Indeed, social-justice research has been effectively eliminated from the land-grant college system.[6]

A major criticism of Goldschmidt's study was that the two towns were of different ages. Arvin developed during the 1920s and 1930s, Dinuba during the 1900s and 1910s. The critics argued that since Arvin was about twenty-five years younger, it had not had time to develop the superior qualities of Dinuba. But Goldschmidt had already demonstrated the inadequacy of this argument in his original study. Adjacent towns such as Delano, Wasco, and Shafter, which grew at about the same time as Arvin but had small-scale producers, had characteristics more similar to Dinuba. In addition, regardless of age, other towns—such as Tipton, Pixley, Buttonwillow, and Mendota—with large-scale farm operations had town characteristics similar to Arvin's.[7]

Arvin and Dinuba Reexamined in the 1970s

In 1977 Steve Peterson, a researcher with the California Department of Housing and Community Development, updated Goldschmidt's study and

TABLE 12.3
Arvin and Dinuba Compared: Town Characteristics, 1944

Characteristic	Large-Scale Farms Arvin	Small-Scale Farms Dinuba
Government	not incorporated; governed by county	incorporated 1906; governed by an elected city council
Elementary Schools		
In Town	1	4
In Rural Areas	$\frac{1-2}{2-3}$	$\frac{4}{8}$
Teacher Turnover (in town only)	14 of 22 did not return for the 1943-1944 session	3 of 22 did not return
Trade and Social Groups	5	15
Family Participation Rates in Group	31%	47%
Churches	7-9	15-17
Church Participation	59%	72%
Retail Businesses	62	141
Gross Retail Sales	$2,535,000	$4,383,000
$ Sales per Capita	$407	$592
Kinds of Retail Sales		
Food	24%	19%
Household Supplies	10	15
Agricultural Equipment and Supplies	12	21
Clothing	4	11
Autos and Supplies	36	23
Liquor	9	7
Miscellaneous	$\frac{4}{100}$	$\frac{5}{100}$
Volume of Business		
Low (under $10,000)	34%	45%
High (over $10,000)	$\frac{66}{100}$	$\frac{55}{100}$
Capital Investment in Businesses		
Low (under $5,000)	81%	69%
High (over $5,000)	$\frac{19}{100}$	$\frac{31}{100}$

Source: See Table 12.1.

Figure 12.3. Family farmers support small towns: the central business district of Stillwater, Minnesota, 1970.

found that indeed age of settlement was not the critical factor, as more years had elapsed since 1944 than separate the towns by age.[8] On the basis of educational, social, and cultural facilities; educational attainments; socioeconomic characteristics; and range of retail businesses the differences found in Arvin and Dinuba in 1944 still held in 1977. The large percentage of agricultural laborers found in Arvin indicated that large-scale farming, with its resulting lower standard of living, continued to predominate. Dinuba, on the other hand, was still dominated by small-scale farmers and had a prosperous commercial business district and a higher standard of living (see Table 12.4).

TABLE 12.4
Arvin and Dinuba Compared, 1977

Characteristic	Large-Scale Farms Arvin	Small-Scale Farms Dinuba
Churches	12	28
Parks	2	5
Playgrounds	0	8
Theaters	1	2
Newspapers	1	2
Libraries	1	1
Radio Stations	0	1
General Hospitals	2	5
Physicians/Surgeons	1	5
Dentists	1	2
Optometrists	1	2
Chiropractors	1	2
Elementary Schools	2	4
Jr. High Schools	0	1
High Schools	1	1
Civic/Social Organizations	7	40
Registered Voter Participation	75%	91%
Retail Businesses	58	126
Total Businesses	96	222
Gross Retail Sales	$12,577,000	$24,181,000
Gross Total Sales	$15,908,000	$27,021,000
Average Farm Parcel	165 acres	24 acres
Median Family Income	$5,903	$7,535
Median School Years Completed	8.3	10.4
Owner-Occupied Housing	52%	62%
Population (1974/1975)	6,013	8,590
Population Doubled	not doubled in 30 years	in less than 25 years
Occupational Structure		
White-collar Workers	36%	54%
Blue-collar Workers	26	32
Farmers and Managers	0.4	0.17
Farm laborers and Foremen	37.6	13.7
	100	100

Source: Small Farm Viability Project, The Family Farm in California, November 1977, pp. 2-13.

130 Towns in the San Joaquin Valley

In 1977, Isao Fujimoto and his rural sociology associates continued the work Goldschmidt was prevented from doing in 1944. They compared 130 towns in the 8 San Joaquin Valley counties to determine the relationship between the control of major agricultural resources—namely, land and water—and the quality of community life.[9] Using the yellow pages of telephone

Selma   Huron

	FARM FACTS		
11	farm owners	1	(1)
(3) 11	resident owners	0	
9	farms	1	(2)
9	resident farmers	0	
7	resident owner-operators	0	
2	resident leasee operators	0	
7	full time farmers	1	
31	people living on land	0	
$916,000	gross value of farm production (4)	$590,000	
$1,092,000	property value (5)	$412,000	
$25,394	property taxes paid	$8,627	

	CITY FACTS	
9,036	population	2,539
287	number of businesses	55
22	manuf. & processing plants	11
0	farm corp. offices	18
$43,317,000	value of retail taxable sales	$7,350,000
1	hospitals	0
6	doctors	1
6	dentists	0
34	churches	2

Figure 12.4. The tale of two square miles: a comparison of scales of farming in the San Joaquin Valley of California. Selma is surrounded by small farms and Huron is surrounded by very large farms. Reprinted by permission of National Land for People, 2348 N. Cornelia, Fresno, California 93711.

directories, the towns were divided into high and low complexity. The presence of a high school represents a critical level of complexity. Towns with a high school tend to be of sufficient size to have other higher-order central place functions, such as dentistry and pet shops. The complexity of a community indicates the range of choices available and thus serves as an index of the town's quality of life. The towns were also divided by their cropping patterns and water systems. Aerial photographs allowed divisions into large-scale farming (cropping patterns of 640 acres or larger) and small-scale farming (cropping patterns of 160 acres or less). Water systems were either democratic or undemocratic: "irrigation district" denoted a democratic jurisdiction, where voting was on a one-person, one-vote basis, and "water district," as in Westlands Water District, denoted an undemocratic district, where voting was based on acreage irrigated or its monetary value (see Figure 12.4).

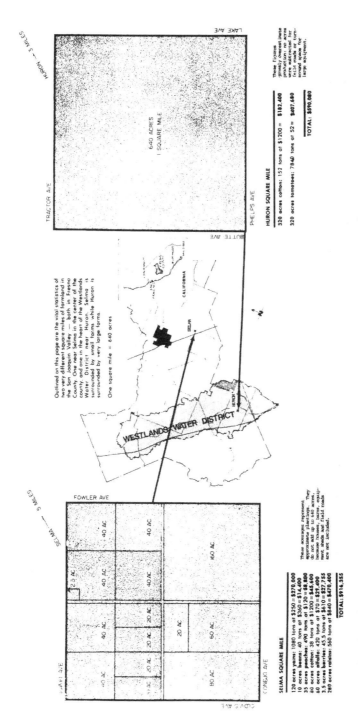

Outlined on this page are the vital statistics of two very different square miles of farmland in the San Joaquin Valley — both in Fresno County. One near Selma in the center of the county and one in the heart of the Westlands Water District near Huron. Selma is surrounded by small farms while Huron is surrounded by very large farms.

One square mile = 640 acres

HURON SQUARE MILE

320 acres cotton: 152 tons at $1200 = **$182,400**

320 acres tomatoes: 7840 tons at 52 = **$407,680**

TOTAL: **$590,080**

These figures generally represent acres in production; no acres were subtracted for field roads or turn-around space for large equipment.

SELMA SQUARE MILE

120 acres yams: 1080 tons at $250 =	**$270,000**
10 acres beans: 40 tons at $360 =	**$14,400**
35 acres peaches: 490 tons at $120 =	**$58,800**
80 acres cotton: 38 tons at $1200 =	**$45,600**
60 acres alfalfa: 420 tons at $70 =	**$29,400**
3.5 acres berries: 45.5 tons at $610 =	**$27,755**
280 acres raisins: 560 tons at $840 =	**$470,400**

TOTAL: **$916,355**

These acreages represent approximately plantings. They do not add up to 640 acres because houses, barns, equipment sheds and field roads are not included.

640 ACRES
1 SQUARE MILE

Figure 12.4 (cont.)

When all three variables were compared—complexity of town services, scale of cropping patterns, and water use—Goldschmidt's original findings were confirmed. Areas with large-scale farming and undemocratic water districts had noticeably fewer towns that provided a smaller range of services. On the other hand, towns associated with small-scale farming had proportionately more elementary schools, dentists, pharmacies, and medical specialists. Table 12.5 provides a summary of services and facilities available in two average towns, one surrounded by large-scale farming and an undemocratic

TABLE 12.5
Consequences of Scale of Production and Types of Water Control in Two Average San Joaquin Valley Towns, 1977

Town A: Large-Scale Farming; Undemocratic Water System

Services	Businesses	Facilities	Clubs
Post Office	none	none	4-H
Elementary School			

Town B: Small-Scale Farming; Democratic Water System

Services	Businesses	Facilities	Clubs
2 Elementary Schools	Department Store	Incorporated City	4-H
Dentist	2 Hardware Stores	Local Government	
Health Clinic	3 Grocery Stores	Community Center	
Pharmacy	Lumberyard		
Agriculture Consultant	2 Beauty Salons		
Farm Labor Office	Funeral Home		
3 Churches	Plumber		
Health Services	Welder		
Social Services	Body & Paint Shop		
Police Department	Oil-Gas Station		
Fire Department	Garage		
Post Office	Parts Store		
P.O. Home Delivery	4 Bars		
Newspaper	2 Cafes		
Town Festival	2 Fast Food Restaurants		
	Movies		
	Processors		
	2 Crop Dusters		
	Farm Equipment		
	Farm Equipment Repair		
	Irrigation System		
	Labor Contractor		
	Dealers/Brokers		
	Bank		

Source: Isao Fujimoto, "The Communities of the San Joaquin Valley," in U.S., Congress, Senate, Priorities in Agricultural Research of the U.S. Department of Agriculture--Appendix, Subcommittee on Administrative Practice and Procedure of the Committee on the Judiciary, 95th Cong., 2d sess., 1978, part 2, p. 1396.

water system, the other surrounded by small-scale farming and a democratic water system.

Both Goldschmidt's and Fujimoto's studies were conducted in California because agrarian capitalism was developed here in its most advanced form. That is, the negative economic and social consequences of large-scale farming subsidized by governmental farm programs, agricultural research, and tax laws were most evident in that state. But if this structural analysis is sound, it should also be verifiable elsewhere in the United States. Researchers in Kansas have provided such verification.

Goldschmidt's Thesis Tested in Kansas

A team of social scientists tested some of Goldschmidt's conclusions in 105 Kansas counties in 1977.[10] Contrary to the Arvin-Dinuba study, the researchers assumed a mutually reinforcing tendency among agricultural structure, social class, and community social well-being. Using factor and regression analysis, they found two patterns. First, although capital-intensive (or large-scale) agriculture resulted in greater total wealth, it contributed to greater inequality of income distribution and to the growth of a dualized

Figure 12.5. Small towns provide critical services to family farmers. This small northern Minnesota town provides grain storage in the elevators along the railroad and agricultural and retail services along the main highway that runs through town.

agricultural structure. That is, a large number of poor agricultural workers and their dependents, frequently seasonal and often migratory, were economically dependent on a few wealthy landowning families. This dependence created inequality not only in the fields but also in the small towns whose prosperity depended upon a wide sharing of the land and the wealth produced from it. Goldschmidt had reached similar conclusions in California thirty years earlier. Their second finding seems to contradict Goldschmidt. "Where we found that a high ratio of wage labor to independent entrepreneurs [farmers] is associated with high average community and individual well-being, Goldschmidt found exactly the opposite."[11]

Both of these contradictory results are consistent with the internal colonialism model, used by these researchers, if one recognizes that the Kansas study reflects an earlier stage of capitalist development of agriculture than was or is true of the Central Valley in California. The penetration of capital in the form of industrial processes and labor relations into agriculture is much more advanced in California and in the irrigated West in general than in Kansas and in most of the rest of the United States. Even in Kansas, the results suggest greater inequalities in counties with strong linkages to industrialized agriculture. As these linkages are intensified by vertically and horizontally integrated national and multinational corporations, the outlines of a dualized agriculture will become sharper and result in a decline in the average well-being of local communities as well as greater individual inequalities. Yet wherever agriculture is in the early phases of being industrialized, and the resulting demise of rural communities is not immediately apparent, family farmers will be particularly susceptible to the myth of the family farm because it camouflages the inevitable long-term structural inequalities and inadequacies and focuses attention, instead, on the shortcomings of individuals. By the time the community consequences become obvious and inescapable, agribusiness is so deeply entrenched that reversing the trend may appear to be an impossible task. It is not impossible, however, and a significant number of people have had the courage to resist.

Part 6

Conclusion:
Agrarian Democracy or
Agrarian Capitalism?

13
Farmers Challenge Agribusiness

Raise less corn and more hell.
—Mary Elizabeth Lease, Kansas Populist

If you want to understand something, just try to change it.
—Urie Bronfenbrenner

Speaking for nineteenth-century capitalists, Henry Adams declared that "a capitalistic system had been adopted, and if it were to run at all, it must be run by capital and by capitalistic methods."[1] But Henry Adams and his fellow capitalists forgot that the United States was also based on democratic principles that directly contradicted capitalism. The Declaration of Independence proclaimed our inalienable rights to life, liberty, and the pursuit of happiness. How could these self-evident truths be realized under a capitalistic system? Presumably, political rights would be used to guarantee economic rights; farm movements would give meaning to Jefferson's agrarian democracy. But, although agrarian capitalists might like to think of *capitalism* as the ideal system for farmers, the tillers of the soil frequently employed *democratic* principles to organize production through cooperatives and nationalization of banks. Agrarianists often viewed the power of capitalists, derived from economic inequality, to be the enemy of democracy.

Capitalists sought to be identified with justice, but this identification has been rejected repeatedly by farmers. Throughout U.S. history farmers, individually and collectively, have challenged the drastic transformations brought on by agribusiness. Many farm organizations have demanded changes—some minor, some major—in the political and economic structure of agricultural institutions.[2] These social movements have tried to force change through educational organizations, to help farmers understand their predicament and organize them to put pressure on the government to aid farmers; through cooperative organizations, to eliminate the profits and manipulations of middle merchants; and through political parties, to seek control of the government in the interest of farm and tenant families. Regardless of the methods used to redress grievances, these farmers have employed grassroots democracy to resist domination from the outside and from the top. Most agrarian movements have been explicitly opposed to big business in general and agribusiness in particular.[3] Historian S. M. Lipset went so far as to say that "the history of American political class consciousness has been primarily a story of agrarian upheavals."[4]

Populism, 1870–1890

As settlers moved west seeking the Jeffersonian ideals of security and freedom, farmers found themselves at the mercy of the railroads, which took their produce to eastern markets, and bankers, who lent them money for equipment and supplies. The Populists—first in Texas, then throughout the South and Great Plains states—and other cooperative farm groups in the Midwest experimented from the 1870s to the 1890s with alternatives to private grain merchants, railroads, farm implement dealers, banks, and the national monetary system. The Populists—blacks and whites, cotton and wheat farmers—faced three choices: (1) become more "efficient" and compete with agribusiness (the conservative view, held today by the Farm Bureau); (2) concentrate on farm cooperatives and bypass agribusiness (the progressive view, held today by the Farmers Union and the National Farmers Organization); or (3) change the structure of society and eliminate agribusiness (the radical view currently held by socialist groups but not by farm organizations). The Populists, with their "socialism without doctrine," to use Lipset's phrase, usually made the third choice.

In the agricultural depression of the 1870s, farm organizations such as the Patrons of Husbandry and the Farmers' Alliance turned from educational programs to political action in their search for basic changes in the economic structure affecting agriculture. Three approaches were used. First, state governments were petitioned to regulate railroad freight rates. Although such regulation is taken for granted today, state regulation of the railroads was initially ruled unconstitutional in various states. The Populists quickly realized that more fundamental changes were needed, so in 1896 the People's Party supported the nationalization of the railroads. Their second attack was aimed at the banks' power to issue credit—the lifeline for farmers even in the 1870s. The Greenback and People's parties demanded that the federal government issue and regulate all money.[5] This radical demand by farmers has become an established practice today. Third, in order to have federal laws enacted to achieve the goals of monetary reform, farmers had to have power in the federal government itself. The Populists realized that they needed urban allies, particularly from radical labor groups in the East. According to Lipset, "This effort by agrarian leaders to find labor allies who would support the demands of the farmers in return for rural support of labor's program is characteristic of American agrarianism."[6]

The threat and potential power of such a rural-urban alliance was verified by Henry Adams's reaction: "Nothing could surpass the nonsensity of trying to run so complex and so concentrated a machine [the U.S. economy] by Southern and Western farmers in grotesque alliance with city day-laborers."[7] The threat was real. By joining forces, farmers and other workers could achieve economic equality and social justice that would be unobtainable for each separately. The Populists "tried through democratic politics to bring the corporate state under popular control."[8] The ideal of grassroots democracy was based on generosity grounded in mass dignity. Populists believed that by working together individuals would become free. Ironically but typically, the democratic ideals of the Populists were labeled insurgent ideas in the very culture that fancied itself democratic.[9]

Populism offers a relevant lesson for today's farmers: Insurgent movements do not result from hard times and do not depend on the property relations of society's members (tenants versus family farms, for example). Rather, they emerge from a sequential set of organizational stages, which Lawrence Goodwyn called "movement culture."[10] " 'Individual self-respect' and 'collective self-confidence' constitute, then, the cultural building blocks of mass democratic politics. Their development permits people to conceive of the idea of acting in self-generated democratic ways—as distinct from passively participating in various hierarchical modes bequeathed by the received culture."[11] The independence of the populists' thinking is reflected in their 1896 demand that the president, vice-president, and U.S. senators be elected directly by the people. But, alas, only one of their democratic demands was achieved; the president and vice-president are still elected by an exclusive club, the electoral college. In the end, populism was as effective and exciting as it was because its programs for social change were based on democratic principles, self-respect, and economic justice.[12]

Progressivism, 1900–1930

The Non-Partisan League carried the traditions of populism into the 1910s and 1920s. This organization had a class program and class strategy to end bank usury and gross exploitation by elevator companies. The league, strongest in North Dakota, gained control of the state legislature in 1918 and passed numerous progressive laws that established the only state-owned bank and grain elevators in the country, a Home Building Association to lend money at low interest rates, a graduated state hail-insurance fund, a workers' compensation fund that assessed employers for its support, the eight-hour day for working women, and regulation of working conditions in the mines.[13] The league had the open support of labor unions and the Socialist Party, but its greatest strength came from wheat farmers in the northern Great Plains (who would organize another protest movement sixty years later in the form of the American Agriculture Movement).

But for the most part the number and impact of radical farm groups declined after 1900. Although progressivism continued the populist tradition of seeking economic change for the benefit of farmers, the progressive movement was very different. It was less direct, far less radical, and not concerned with the most disadvantaged farm and urban classes. Indeed, "the Progressive philosophy turned invariably to the model of vertical mobility as a solution for individual farm operators."[14] Populists sought structural solutions to rural and urban societal problems, but Progressives sought government assistance for individual solutions to farm problems. It was inevitable that such policies would help economically more successful farmers. The basic and deep-seated maladjustments in rural areas were not addressed, and the solution for "marginal" and tenant farmers was to move them out of agriculture. The ideological roots of current federal agricultural policies date back to the progressive era.

The most conservative farm organization, the Farm Bureau, emerged during this period. With the passage of the Smith-Lever Act of 1914 the federal government made matching grants to the states to support county agents. According to Samuel Berger, "The county agent gradually became the publicly paid organizer of the Farm Bureau."[15] Typical of the progressive philosophy, the Farm Bureau Extension Service was an organization of and for wealthy farmers, the only ones who could afford to experiment with new agricultural techniques and pay the annual dues. By 1919 the bureau's class bias was so well known that "the U.S. Secretary of Agriculture, David Franklin Houston, called upon farmers to join the Farm Bureau in order to 'stop bolshevism.'"[16] In Berger's words, the bureau was "the right wing in overalls," and its clout in government was revealed in the destruction of the next wave of farm radicalism.

The New Deal

During the Great Depression, radical farm organizations, among them the Farm Holiday movement, called for direct action by farmers to withhold produce and to stage a general strike.[17] At farm foreclosures, bank officials were turned away by angry mobs of farmers, and national guards were brought in to allow farm auctions to take place (see Figure 13.1). In response to this rising wave of radicalism, Congress passed the Agricultural Adjustment Act of 1933, which was strongly supported by all of the major traditional farm groups. The act gave the secretary of agriculture the power to raise prices by reducing output. The initial plan was for the Agricultural Administration Agency (AAA), which administered the act, to be a strong central agency to counter the strong political and economic clout of wealthy farmers at the local level. But the conservative, proagribusiness Farm Bureau, through its governmental contacts, had the AAA's power decentralized to the county extension service level. The bureau's unique relation to the extension service allowed it to benefit from this massive new decentralized government program. Indeed, "the periods of greatest government involvement in agriculture have been the periods of the Farm Bureau's greatest growth."[18]

After the passage of farm legislation to deal with the immediate farm crisis, President Franklin Roosevelt and his New Deal administration sought more fundamental social reforms. But, having grown large and powerful because of government programs, the Farm Bureau opposed deep-seated rural changes. While espousing the interests of family farmers, the bureau served the interests of agrarian capitalists by attacking progressive farm legislation, such as that establishing the Farm Security Administration (FSA).

Figure 13.1. Banks control agriculture. The government's police power is used to support owners of capital and property, not those who work the land: at a 1933 farm foreclosure sale, the National Guard is used to control an angry group of Iowa farmers. *Milwaukee Journal* and USDA photo.

The FSA, launched in 1937, was the poor families' Department of Agriculture and the most extensive effort by the federal government to deal with the problems of the rural poor, subsistence farmers, and rural blacks in the South. Its major focus was on rural rehabilitation, which was based on the philosophy that it is better to make needy families self-supporting than to give them mere relief.[19] The FSA provided loans to farmers and cooperatives, operated a chain of migratory labor camps in the West, resettled unemployed urban families to rural areas, and helped poor and black cotton farmers in the South plant gardens and shift to dairying for higher incomes.

This progressive farm program, which by no means attacked the structural causes of rural poverty and low farm incomes, was destroyed by the Farm Bureau, the largest farm organization, in the name of the best interests of

Figure 13.2. Alternative agricultural organization. (Top) During the 1930s, the Farm Security Administration (FSA) made low-interest loans available to farmers. A large Arkansas family bought this 40 acres in California for $1,000 in 1938. Dorothea Lange, FSA photo. (Bottom) The Resettlement Program of the FSA settled urban families on the land. Near Thief River Falls, Minnesota, one-and-a-half-story houses and small, gothic-roofed barns are characteristic of these self-sufficient settlements. Most families settled on 160 acres of marginal land, usually bought for $500.

farmers![20] The FSA was attacked by agribusiness interests, inside and outside Congress, and its resettlement projects were compared to Soviet *kolkozes*. Diversification of agriculture in the South irritated southern cotton landlords and commercial dairy interests in the North. Loans for cooperatives to buy grain elevators increased the hostility of large private grain dealers in Minneapolis. The FSA's migrant camps alleviated the degrading living conditions, but displeased growers. The FSA tried to introduce cooperative farming and arranged for ninety-nine–year leases to keep the cost of farming within reach of low-income families; agribusiness proponents cited these leases as contempt for the concept of fee-simple ownership of land. In short, the FSA programs represented a threat to agrarian capitalism. In the end, Congress accepted the arguments made in support of private property rights and large-scale growers in the South and West and rejected the programs that served the interests of small farmers and rural poor families. In 1946 the FSA was terminated and some of its functions were transferred to the Farmers Home Administration, whose perspective and funding policies reflect mainstream agricultural policy—that is, the perspective of agribusiness. The dominance of agribusiness was once again confirmed by Congress.

Farm Protest Movements in the 1970s

The agricultural price boom of 1973–1974 led to two years of high farm prices. After the peak in August 1973, farm prices fell 57 percent while farm costs increased 33 percent.[21] This newest price-cost squeeze threatened the financial survival of many farmers. The USDA estimated that as many as 10 percent of the farmers in the nine-state wheat region were forced off the land by low prices in 1977. Another 40 percent had to refinance their operations.[22]

American Agriculture Movement

In response to these conditions, a national farmers' strike began in Springfield, Colorado, in September 1977 and spread to thirty states by December 1978. The strikers refused to plant, harvest, or buy nonessential goods "until the federal government stopped their slide into bankruptcy." They staged numerous tractorcades at state capitols and in Washington, D.C. By late 1978 the American Agriculture Movement (AAM) shifted from strikes and tractorcades to educational and legislative action to "get farmers aroused so they will write to their congressmen and senators in support of a better farm bill."[23]

AAM had four demands: (1) agricultural products should be sold at 100 percent parity; (2) an agricultural producer board should be created to review federal agricultural policy; (3) restrictions should be placed on agricultural imports; and (4) federal policy announcements relating to agriculture should be made far enough in advance to permit farmers to plan their operations. In Catherine Lerza's words, "In essence, farmers wanted protection from uncertainties of the free market, without giving up the freedom of that free market."[24]

The conservative nature of the AAM farmers is illustrated by their stand

Figure 13.3. Protesting farmers in the 1930s. Farmers dump milk to protest low farm prices. Such tactics continue to be used particularly by the National Farmers Organization. FSA photo.

Figure 13.4. Protesting farmers in 1979. Members of the American Agriculture Movement drive through the reflecting pool in front of the Capitol in Washington, D.C. Photo by Joel Richardson, *Washington Post*. Reprinted by permission of the publisher.

on parity and how to achieve it. The AAM wanted the profits of commodity dealers and other middle merchants, who now "get more than their fair share," to be cut by a minimum farm-price law, which would make it illegal to buy or sell farm products at less than 100 percent of parity.[25] Parity was defined by the strikers as including the costs of production plus a "reasonable return for our investment and labor—about eight percent." To achieve their goal, strikers wanted the federal government to raise the price floor of its loan-support to parity levels. Taxpayers would subsidize the price of wheat, for example, by nearly $3 a bushel. Strikers did not want to receive subsidies directly but instead wanted the government to maintain parity-level "free-market" prices with crop restrictions that would prevent the cheapening effects of overproduction. Consumers would, of course, continue to pay higher food prices. Instead of direct subsidies that would be part of the federal budget, the AAM wanted hidden subsidies passing for free-market solutions.

Who would benefit from the 100 percent parity payments, whether direct or hidden? Farmers in the $40,000 and over agricultural sales category have received more than 60 percent of all federal farm payments, with more than one-third of total payments going to the $100,000 and over farmers. In 1977, big farms received about $2,340 each in federal payments; middle-sized farms received about $500 each.[26] The larger the farm, the more beneficial parity pricing would be.[27] Richard Margolis characterized the AAM as "middle-sized farmers, who ought to know better, [who] volunteer to do the work of the rich, and at no extra charge."[28]

Farm Women's Organizations

Female farmers for the first time started their own general farm organizations in the 1970s to achieve independence from male-dominated farm organizations and financial security in agriculture. The American Agri-Women (AAW) was formed in November 1974 as a coalition of several state organizations with 3,000 members in thirty-four states. In addition, Women Involved in Farm Economics (WIFE) was established in December 1976 and by 1979 consisted of 116 chapters in eighteen states. Like their counterparts in the antiwar movement, farm women realized that their work in traditional farm organizations and in the American Agriculture Movement was not recognized as equal by men. Farm women also learned that male farmers were not willing to fight for matters that affect mostly women, such as estate-tax laws that excluded them as farm owners. As "good farm wives," they shared the responsibilities and work on farms but not the fruits of their labor when their husbands died or they were divorced. Ironically, farm women's organizations distinguish themselves from the feminist movement, yet their goals and actions are identical to those of feminist organizations.

Farm protest organizations in the 1970s, whether female- or male-run, identified with propertied classes even when their economic interests necessitated alliances with other working people. In 1976 AAW adopted as a major goal "to create public acceptance that farming is a business that must make a profit to survive." The organization actively opposes anything it perceives as a threat to the profit potential.[29] These threats include state and federal laws related to minimum farm wages, working conditions and housing of agricultural workers, regulations on the use of herbicides and pesticides, and any consumer movements aimed at keeping food prices down. The farm women's movement wants the family farm to survive, but its policies exclude Chicano farm workers, black sharecroppers, and even young, white organic and subsistence farmers, and yet embrace the short- and long-term interests of agribusiness. For the most part, the farm movements to date have failed to address the structural problems of agriculture in the U.S. economy. Their emphasis has been on symptoms that reappear in new forms generation after generation, sapping the protesting energy of the remaining farm families while achieving no permanent solutions. Indeed, these movements have only delayed the "inevitable"—the demise of family farming.

Farmers have struck randomly at the most visible economic evils that have affected them. They have developed agrarian but not class consciousness. To use an earlier-cited metaphor, they see the islands in the sea but fail to understand that all the islands are joined together under the reflecting surface of the ocean. In the past, they opposed banks, railroads, wheat elevator companies, inflation, and government policies, "but they saw each evil as an evil in itself, not as part of the total economic system," known as monopoly

Figure 13.5. Rugged individualism on the farm. An Iowa farm family proclaims its independence of government subsidies—but ignores its dependence on and exploitation by agribusiness firms.

capitalism.[30] Family farmers, agricultural workers, and other producing classes can no longer afford to define themselves within the framework of inherited power relationships, which have manipulated farmers and the public into supporting the political status quo of the corporate state.[31] Understanding the structural class relations of farmers is therefore essential to the success of any future challenge to agribusiness.

14
Where Have All
the Family Farmers Gone?
A Structural Analysis
of U.S. Agriculture

The myth of family farming obscures the dominance of agribusiness and the root causes of rural problems. The problems, discussed throughout this book and briefly summarized here, are numerous and wide ranging.

1. Privately owned rural land is highly concentrated.
2. The concentration of land into large-scale and industrial farms has been facilitated by government policies that were purported to support family farmers, for example, nineteenth- and early-twentieth-century land grants and the Reclamation Act of 1902.
3. Large-scale farms are more profitable than family farms, even though they are no more efficient.
4. Family farmers are heavily in debt for real estate and other productive resources.
5. Family farmers must rely on off-farm employment to survive.
6. Contract farming has eroded the independence of family farmers, who are now effectively employees of agribusiness.
7. Family farmers are at the mercy of farm oligopolies and oligopsonies, which control the prices farmers must pay and the prices they will receive.
8. Agribusiness has the advantages of horizontal and vertical integration.
9. Federal tax laws result in higher land prices, overproduction of certain crops, expansion of larger-than-efficient farm sizes, absentee ownership, and the demise of family farms.
10. Federal farm programs disproportionately benefit large-scale farmers in terms of their numbers and income. These programs reward the wealthy for not working and penalize the rural poor for not having land to work.
11. Research at land-grant colleges and universities subsidizes mechanization of crops on large-scale farms and displaces agricultural workers without compensation.
12. Federal farm programs discriminate against blacks.
13. Black land-grant colleges have been unequal to comparable white institutions.

14. Blacks, who never owned much farmland, are experiencing a disproportionate decline in landownership throughout the South.
15. Seasonal farm families are the most exploited, having the poorest standard of living and working conditions. Agribusiness benefits from this exploitation.
16. Growers have used racism to divide migrants. This division has made unionization difficult.
17. Out of economic necessity, family farmers and especially migrants are forced to exploit their children's labor on farms.
18. Large-scale farms and integrated agribusiness have destroyed agriculturally dependent towns and have reduced the quality of life of those that still exist.
19. The myths that support agribusiness have been used to steer migrants, tenants, and family farmers away from fundamental, enduring change and toward immediate, piecemeal solutions.

To understand the source of these rural problems we must ask: Are the problems haphazard misfortunes, or are they structurally related to each other in a significant, causal way? Who benefits from our current agricultural system and who pays the price? To solve these problems, we must further ask: Will people be rewarded for merely owning wealth or for working to produce wealth? Will investment decisions be made by a few people to benefit themselves; by a majority of people to benefit a few; or by a majority of people who work and are collectively rewarded for their skills, risks, and labor? Will human material needs, such as food, shelter, and health care, be as basic as civil rights? The answers to these questions will mold future developments in rural America. In agriculture the ideology of laissez-faire economics has been used to mystify the tyranny of economic exploitation, which is no longer accepted by workers in most other sectors of the economy. Farming plays a special lingering role in the perpetuation of the myth that political democracy and capitalism are compatible. At the same time, farming represents the potential to challenge that same myth to create an egalitarian future.

For conservatives, the importance of agribusiness in U.S. agriculture is the inevitable and desirable result of the natural workings of a free market. Those leaving agriculture—family farmers, tenants, migrants, and small town merchants—are failing through their own "inefficiency." The system is merely "removing unnecessary fat." The treatment of minorities in agriculture is justified by their "laziness" and "racial inferiority."

Liberals recognize the fallacy of the conservative argument. But their reform attempts to pass legislation setting upper limits on agricultural payments per grower; improving health, education, and shelter for migrants; and providing low-interest loans to farmers, for example, have not prevented the demise of family farming and the exploitation of racial minorities in rural areas. Liberals have sympathized with the victims of agribusiness, but their analysis has not recognized the structural relationships of problems under agrarian capitalism. The social and economic problems we face today are inherent and inevitable results of capitalism.

Capitalism is based on inequality, democracy on equality. The U.S. constitution guarantees life, liberty, and the pursuit of happiness, and under this political democracy individuals are treated equally (one vote per person and equal rights for all). An economy based on labor contributions is also based on equality because each member of that society can contribute labor. But an economy based on capital accumulation for its own sake, as capitalism is, creates a society of economic inequality because access to productive resources (land, factories, money) is not equally available. Under capitalism, "equality of opportunity"—espoused in the land grants of the nineteenth century but not labeled as such until the 1960s—is more concerned with opportunity than with equality. Individuals who fail to achieve equality can then be blamed for not having taken advantage of opportunity. In the United States political rights are nominally guaranteed, but economic rights must be fought for repeatedly. Our current economy makes individual farmers responsible for their own economic survival under impersonal market forces. But explicit recognition of agrarian capitalism—the system under which U.S. farmers work—compels an analysis of the problems facing family farmers from a different perspective, one that finds the nature of the market economy itself inadequate.

A two-sided or class analysis of U.S. society exposes the relationship between persons with capital (capitalists) and persons with labor (workers). Family farmers, migrants, and other workers earn their income from their labor, while capitalists, including the owners of agribusiness, receive their income because they own the workplaces—land, machines, and buildings—in which workers toil. In its most abstract form the struggle is between labor and capital, between those who produce wealth and those who own and appropriate wealth. Under capitalism the owners of the means of production decide who, how many, and under what conditions employees will labor. Workers respond by demanding control of their workplaces and the decisions of how and for what purpose the value of their labor shall be used. Labor and capital, then, are fundamentally opposed to each other and constitute the oppositional unity of agrarian capitalism. Such a class analysis is both more comprehensive and more powerful than conventional social science analysis. Insisting that no classes exist in the United States is the best cloak for class domination, just as idealizing mythical family farms disguises agribusiness domination. In rural areas, the different perspectives of capital and labor lead to class conflict and struggle: agribusiness attempts to impose its social order, and family farmers attempt to assert their autonomous rights.

These rural class relations, however, are more complex than a simple polarization of capital and labor. Large-scale producers such as larger-than-family farmers, industrial farmers, and corporations in farming represent the purest form of agrarian capitalists, the agrarian bourgeoisie; hired farm workers clearly represent labor, or the rural proletariat. But small-scale producers such as family and tenant farmers and sharecroppers are caught somewhere in between, having characteristics of both the capitalists and the proletariat; they are the protoproletariat. All three classes have different control over resources, decision-making processes, and labor relations. Roger

Barta, a rural sociologist, has pointed out that as long as the agrarian structure is dominated by the capitalist market, the inevitable tendency is toward a deeper differentiation or polarization of the rural population and the proletarianization and pauperization of family farmers.[1] The term protoproletariat, "a group clinging to a semantic grey zone,"[2] is especially appropriate for family farmers. Although the myth of the independent, self-determining family farm allows family farmers to identify with the agrarian bourgeoisie, their structural relations tie them economically to the rural proletariat. Their protoproletariat status distinguishes them from landless farm workers yet recognizes their overwhelming similarity to them. With the continuing penetration of capital into the countryside—whether through greater bank control, contract farming, or direct farm production—the remaining family farmers will first become more marginalized and finally be reduced to the ranks of the rural or, more likely, urban proletariat. Since 1920, almost 50 million farm people have experienced this consequence of agribusiness.

Labor Theory of Value

In the nineteenth century, political economists of the right, such as Adam Smith, and of the left, such as Karl Marx, agreed that wealth is created when human labor transforms nature. That is, the value of a house is determined by the amount of labor expended to cut the trees, saw the boards, lay the bricks, and hang the doors. Even a U.S. president, Abraham Lincoln, understood that "labor is prior to and independent of capital. Capital is only the fruit of labor and would never have existed if labor had not first existed. Labor is the superior of capital and deserves much the higher consideration."

The labor theory of value exposes the inadequacy of agrarian capitalism and hence the myth of the family farm. According to this theory, humans produce wealth with their labor by using natural resources to make useful products. A "good investment" does not, as some people believe, create wealth; rather, it expropriates the wealth created by workers. *Labor is the only creator of exchangeable value.* Human labor is so productive that more wealth is generated than is required for the subsistence of the worker and for raising the next generation of workers. In a subsistence agricultural economy, all wealth or capital is owned by the laborers who produce it. But under agrarian capitalism, the wealth produced by farmers' labor is extracted from them by taxes, usury, rent, and profit. T. Wharton Collens's 1876 illustration of workers' relationship to capital is applicable to farmers' exploitation today (see Figure 14.1).[3]

Taxes are the contributions of labor to support government for social and national purposes. In Figure 14.1 taxes are represented by the sovereign sitting on a throne and holding a book and sword. The book justifies and the sword enforces agrarian capitalism. In the United States, the book is the constitution, which promises justice, yet the state uses taxes, collected from labor, to assist capitalists to dominate and exploit labor even more. Since taxes could be collected and distributed equitably, they would not have to represent exploitation of family farmers, but under agrarian capitalism they inevitably

Figure 14.1. The labor theory of value. Capitalists expropriate the wealth produced by labor from nature. *Source:* T. Wharton Collens, *The Eden of Labor* (Philadelphia: Henry Baird, 1876).

do. The word "monopoly" under the throne is appropriate because the U.S. government preaches freedom and justice but cultivates economic monopoly. The land-grant college system, for instance, is based on taxes, which are used to accelerate the rate of mechanizing agriculture to the benefit of agribusiness and to the detriment of migrants and family farmers. Likewise, the benefits family farmers received from federal land-granting policies were "drowned out in an ocean of abuse and fraud."

To continue the imagery of Collens's illustration, *usury* is the razor that denudes the head of labor of every vestige of hair that the scissors of taxes have left. By charging interest on borrowed money, capitalists, including bankers, extract wealth from farmers. Usury laws depend on the government for their enactment and enforcement. The benefits of usury, Collens said, are "a life of ease, idleness, and unlimited indulgence."

Landlords derive *rent* by allowing others (tenants) to work the land. Rent also occurs as disguised income, which owners of farms receive from the work of hired agricultural workers. Regardless of how rent is collected, the nature and source of this revenue is essentially the same: wealth is derived from the labor of others, not from the labor of landlords. Private property is the key to exploiting labor through rent.

Profit is the value of exchange received in excess of the cost of the products and services exchanged and the average wage necessary for the exchange. Grain merchants, for example, might sell wheat at twice the price they paid for it three months earlier. The distinguishing trait of profit is twofold: buying and selling, or the double act of exchange. Profit is represented in Figure 14.1 as different from usury and rent: the latter are reclining or seated, but profit is standing. Profit is necessarily busy and active: it follows commerce, speculates on demand and supply, and spreads its web over the chances of speculative transactions. These are the activities at which integrated agribusiness firms excel.

In Collens's illustration, taxes, usury, rent, and profit are shown as separate objects being manipulated by different people. In real life, farmers, particularly family farmers, receive income from several sources: *rents* as landlords, *wages* as farm and off-farm workers, government subsidies from *taxes*, *usury* from bank accounts, and *profits* from speculating on the price of agricultural products held in their own storage or on the futures commodity market. The contradictory multiple roles family farmers play conceal what is actually happening to them. Karl Marx and Friedrich Engels described the spiral of downward mobility.

> The lower strata of the middle class—the small tradespeople, shopkeepers, and retired tradesmen generally, the handicraftsmen and peasants [including most farmers]—all these sink gradually into the proletariat, partly because their diminutive capital does not suffice for the scale on which modern industry is carried on, and is swamped in the competition with the large capitalists, partly because their specialized skill is rendered worthless by new methods of production.[4]

Control over the application of new agricultural technologies for saving labor enables the appropriating class to dispense with a still greater portion of the laboring class—migrants, tenants, and family farmers. With greater output and fewer remaining farmers, agribusiness has increased the absolute value of agricultural products, but the relative share that farmers receive of the total food value has declined. Family farmers have had to adapt to capital-intensive, not labor-intensive, technology or be forced out of farming. By identifying primarily with their role as appropriators rather than creators of wealth, most of the remaining farmers have internalized the capitalist perspective and blame themselves and the state for their difficulties rather than agrarian capitalism itself.

Surplus Value

The impulse for agricultural mechanization comes from the drive of capitalists, both urban and agrarian, to increase profits. In its practice, the land-grant college system largely shares this perspective, but in its rhetoric it sings an egalitarian tune.[5] Increased productivity benefits the owners of production (capitalists and landlords) more than the workers, whose labor produces this increased wealth. This contradiction is explained by the Marxian concept of *surplus value*. In an eight-hour day, workers produce enough wealth to pay for their wages in, say, four hours of work. The value they produce in the next four hours goes to the capitalists in the form of surplus value, which in conventional terms is profit. Under these social relations, capitalists get richer for doing nothing and workers receive nothing for working. More surplus value can be created in two ways. (1) An increase in *absolute surplus value*, obtained by lengthening the working day without an increase in wages for the workers (a common practice in the nineteenth century), is unnecessary today because technological developments allow for an increase of relative surplus value. (2) *Relative surplus value* results from increased productivity of labor without any overall reduction in the working day. Workers work fewer hours to produce their own wages and are thus working more hours for their bosses (see Figure 14.2). With farm productivity increasing from an index of 53 in 1910 to 113 in 1975 (100=1967), and with farmers receiving a smaller share of the total food value, relative surplus value has increased in U.S. agriculture.

Among hired agricultural workers, the extraction of surplus value is similar to that of other workers, since they all work directly for capitalists. On family farms the relevance of the surplus value concept is less obvious because farmers work for themselves. They have no direct bosses. Yet surplus value is extracted indirectly from them by agribusiness—through lower prices for products and higher prices for supplies, machinery, and capital—and by the government—through low farm price supports, exemptions from labor legislation, and agricultural research to increase productivity without assuring that the benefits will go to working farmers. For family farmers as well as hired farm labor, their exploitation can be defined as the amount of surplus value produced by their free labor. The expropriation of their labor and surplus value by agribusiness represents class exploitation.

But the accelerated extraction of relative surplus value from farmers has resulted in an obfuscation of their true exploitation and their common social relations with other exploited workers. U.S. consumers have been told by a proud USDA that fewer farmers are producing more. The resulting rise in

Figure 14.2. The exploitation of labor through the extraction of surplus value. Reprinted by permission of Fred Wright, United Electrical, Radio & Machine Workers of America (UE).

absolute net farm income for the remaining farmers and the concomitant increase in food costs disguise the declining share that farmers receive of the food dollar and encourage urban ignorance of the increasing exploitation of the aggregate agricultural sector. The cost of that exploitation is displaced to urban workers in several ways. The migration of millions of farm families, forced out of agriculture and into cities, transfers the costs of rural capitalism to the cities. Furthermore, the extraction of surplus value from farmers allows greater extraction of surplus value from other workers in turn, because they can be paid less for their subsistence needs. Strongly unionized urban workers in the United States have commonly received increased relative wages, which helps to maintain animosity between industrial workers and farmers.

Contradictory Roles of Family Farmers

The family farm myth is accepted by U.S. farmers because they play various conflicting roles that partially confirm the myth (see Table 14.1). From the land they own, farmers as landlords expect to receive *rent*. From the capital they borrow and pay interest on, farmers as capitalists expect to make a *profit*. From their labor, farmers as workers expect to receive *wages*. As owners of land, capital, and labor, family farmers have three distinctive sources of income and these sources represent three conflicting social classes. Each class wants to increase its share of the total wealth produced by labor, and therefore each class stands in conflict with the others.

Although landlords and capitalists are similar because they both control different aspects (land and capital) of the means of production, they stand in conflict as well. Higher land rents for landlords mean less surplus value for farm machinery dealers and grain elevator operators, for example. As capitalists, male farmers reap *surplus value* from hired labor, "free" family labor (women and children), or both, and as landlords, they reap *ground rent* from their ownership of farmland. Ground rent consists of two parts: absolute and differential. *Absolute rent* exists for every parcel of land, regardless of how poor its fertility or how remote its location. *Differential rent* is generated by special advantages such as fertility or proximity to markets.[6] "In the final analysis, ground rent constitutes a transfer of surplus value from capitalists to landowners."[7] Because each family farmer represents both landlord and capitalist classes, this conflict is internalized and hidden within farm families.

TABLE 14.1
Contradictory Roles of Family Farmers under Agrarian Capitalism

Owners of	Sources of Income	Social Classes
Labor Power	Wages	Laborers
Land	Ground Rent	Landlords
Capital	Profits	Capitalists

As a result of these two class interests, which other workers do not have, family farmers tend to identify with capital more than with labor. Historically, this class position has been supported by the distribution of income going to capital and labor. In 1949 labor was receiving only 43 percent of the income in agriculture. By 1968 the relative share of income going to capital in agriculture was 80 percent, with only 20 percent going to labor. This is a complete reversal of relative shares of national income going to labor and capital for the economy as a whole.[8] Agribusiness and the land-grant colleges have encouraged mechanization and other capital-intensive substitutions for labor as necessary and desirable. As a result, in 1976 the average capital invested per farm worker was over $88,000, or 4.5 times the national average of $20,000 per worker.[9] Is it any wonder that farmers identify with capital and not labor?

Access to and control of land in agriculture give rural class relations their unique character (see Table 14.2). As landlords, family farmers associate themselves with the ideals of agrarian capitalism. Many, perhaps most, farmers have earned much more from appreciation in the value of their real estate than they have from crops and livestock production on their farms. Their role as landlords has gained them more than their labor! In a capitalist society, private property rights and capital investment for profit maximization are rewarded more than working for a living. Landownership thus ties family farmers to other proprietary classes and obscures their class relations to workers.[10]

Family farmers thus have the cumulative advantages of labor power, citizenship, and landownership. Tenants and sharecroppers are characterized by having less control over land than family farmers. Their lease or verbal agreements are the basis for their precarious production decisions. Traditionally, these two farm types have been considered less desirable than full ownership because their tenure is not secure and landlords live off their toil. Because permanent and seasonal farm workers are landless, they have only their labor power to sell and are at the mercy of landowners, including

TABLE 14.2
Access to Land Determines Level of Farm Labor Exploitation

	Access to Land	Distinctive Source of Power	Farm Group
Exploitation increases	Control over Land	Landownership	Family Farmers
		Land Lease	Tenants Sharecroppers
	Landless	Citizenship	Permanent Farm Workers Seasonal Farm Workers Migrant Farm Workers
		Labor Only	Undocumented Farm Workers

family farmers. Still, their U.S. citizenship shields them from even more exploitation; undocumented or illegal alien farm workers, whose only value is their labor power, are the most exploited. They have no rights as citizens and control no access to land, so their living conditions and wages are the worst of all agricultural workers and reflect their powerlessness.

This variation in land-based power distinguishes family farmers from the true rural proletariat. But the economic advantages gained by family farmers as landlords and capitalists are continually depleted by agribusiness's expropriation of the large surplus value they produce as workers. This conflict of class interests between agribusiness and the worker role of farmers, and within the family farm itself, inevitably prevents family farmers from profiting as true capitalists. What they gain as apparent profit is extracted from the surplus value of their own labor.[11] Recognition of this contradiction leads large-scale and industrial farmers to eliminate labor from their roles. Surplus value is thus extracted from hired workers, and profit is augmented rather than diminished by exploitation of labor—because someone else's labor, not the farmer's, is being exploited. Under agrarian capitalism, farmers who continue to operate under all three roles (that is, true family farmers) cannot possibly compete with farmers who have externalized the exploited laboring role.

Family farmers may control land, but in today's agriculture many other factors must also be controlled before farmers can be said to control the means of production. Although they control the basic ingredients of land and labor, their lack of control over capital, technology, and innovations renders them powerless as capitalists and landlords. Their actual dependence on input and purchasing agrifirms makes them essentially workers with the appearance of being owners. Although farm productivity per worker outstrips industrial productivity several times, farmers' income is below that of industrial workers. American farmers have been, in the words of political economist James Iron, "squeezed, bled, induced to work hard for little."[12] Instead of believing what was actually happening to them and rebelling against it, farmers reacted to their impoverishment by redoubling their efforts and working harder. Why? In the nineteenth century, American farmers were led to believe that their economic squeeze was attributable to unknowable causes. Today the state plays such a vital role in agriculture that farmers can blame the government for their problems and at the same time demand its protection against agribusiness. Conveniently for agribusiness, the state is blamed for problems originating in the market economy itself, and federal Band-Aid solutions soften the consequences of agribusiness. Also, because family farmers are owners— mortgage-burdened owners—of their own farms, they have a stake in agrarian capitalism and, indeed, something to lose if they do not work hard. The paradoxical incentives to defend the very system that is exploiting the remaining family farmers and has already displaced 50 million farm people gives proponents of capitalism a powerful propaganda advantage.

Because communist countries do not conceal the method by which they squeeze the farmers, and because they do not use private ownership as a means of, in effect, letting selfishness defeat itself, they make themselves look bad. Pointing to communism's collective farms and damning them as examples of "totalitarian slavery," we Americans are able at the same time to point to our equally remorseless system and proudly refer to it as "glorious free enterprise."[13]

In Table 14.3, the various types of farmer roles are compared. Corporations in farming are the purest form of agrarian capitalism. Managers, working for large-scale producers, make everyday decisions and oversee a hired labor force. The ideal of family farmers, in contrast, is that they provide and perform all three functions—the perfect example of self-reliance and independence. In reality, family farmers are only nominal resource owners whose long-term investment decisions are controlled by agribusiness. They make mainly everyday decisions and are predominantly workers: their labor power is their major asset.

Karl Marx's description of the nineteenth-century French peasantry provides insight into the situation of U.S. family farmers in the late twentieth century: "The small-holding of the peasant is now only the pretext that allows the capitalist to draw profits, interest, and rent from the soil, while leaving it to the tiller of the soil himself to see how he can extract his wages."[14] Even

TABLE 14.3
Types of Farmers and Their Economic Roles

Type of Farmer	Capitalist (Resource Owner)	Manager (Decision Maker)	Worker (Labor Power)
Corporation in Farming	X	--	--
Farm Manager	--	X	--
Family Farmer			
Ideal	X	X	X
Actual	x	x	X
Husbands	x	x	X
Wives	at discretion	at discretion	X
Practically	of husband	of husband	X
Legally	--	--	X
Children	--	--	X
Large-Scale Farmer	X	x	x
Small-Scale Farmer	x	x	X
Contract Farmer	x	--	X
Tenant	x	x	X
Sharecropper	--	almost none	X
Hired Farm Worker	--	--	X

X Major Role   x Minor Role

though family farmers work their own land, they are manipulated by agribusiness through its control of processing, storage facilities, transportation, and finance capital. "The small holder can therefore be considered a worker in an agricultural system controlled by urban financial interests."[15]

Within farm families, the roles of men and women are differentiated. Only single female farm owners are treated with the same status as male farmers. In the traditional family (having both husband and wife) farms are generally considered family property, but in most states they are legally the personal property of males. Practically, most farm women perform the same roles as farm men: they do chores, keep records, and make investment decisions. In most states, however, the legal status of farm wives is that of "free" labor only. That is, farm women frequently have no claims on farm assets—which they helped create—when they get divorces (see Box 14.1) or when their partners die. Not all female farmers contribute directly to farm production, but they assume almost complete responsibility for reproduction. The reproduction of farm labor and labor in general (that is, raising children) is conveniently provided by the family, where its cost is hidden and women's labor contributions can be discounted. This dismissal of women's labor is further revealed and reinforced by USDA data in which women's labor contributions equal only one fourth of men's.[16] Failure to consider household labor equal to field labor is typical not only of a sexist society but also of a capitalist economy. Farm children are also legally reduced to "free" labor. In everyday life, women and children often perform other functions, but their roles are totally dependent on the good will of male farmers. (In parallel fashion, factory workers may be given management duties but these responsibilities are given to them [and can be taken away] by management and the plant owners and are not the secure rights of the workers.)

Farmers are both exploited and exploiting. Caught between oligopolistic agribusiness, which supplies agricultural inputs and buys farm outputs, family farmers are at the mercy of market forces beyond their control and benefit. To survive they must reduce their costs. Labor costs are the only productive factor over which they have control; hence, "free" family labor must be exploited. The slogan, "farming as a way of life," persuades farmers to accept their contradictory roles. Outwardly, they deal with the market economy as any other group, but within the family, farmers depend upon the social cohesion and continuity of familial relations to obtain unpaid workers. Because most farm wives and children are unwaged, their relations to capital are hidden; they appear to stand only in a private relation to a male wage earner.[17] Furthermore, farm men exploit their own labor when they confuse wages with profits.

Family farmers relying almost exclusively on "free" family labor lack the more obvious class antagonisms found in large-scale, industrial-type farming. With class tension and exploitation hidden in the family, romantics extol the virtues of small-scale producers and add yet another dimension to the myth of the family farm. Under conditions of agrarian capitalism, conservatives overtly support agribusiness, liberals champion family farmers, and radicals see agribusiness and family farmers as representing opposing classes and per-

---

BOX 14.1. FARM WIVES AND SEXIST EXPLOITATION

Irene and Jim Murdoch were married in 1943. She owned two horses; he had 30 horses and eight head of cattle. They hired themselves out as ranch hands for several years, saved money, and then invested in a dude ranch. For many years they worked side by side. They saved more money and accumulated more property. She did all the household tasks plus "haying, raking, mowing, driving trucks and a tractor, dehorning, vaccinating, branding"—any work that needed to be done.

For five months a year, while Jim was away from the ranch attending to his duties as a member of the forest service, Irene managed the property alone. To their joint ventures, she also contributed money that she personally earned. In 1968, Jim asked Irene to release her dower rights[a] to a piece of their property so that he could sell it. When she refused, he beat her so severely that she was hospitalized. This led to their separation after 25 years of marriage.

Irene Murdoch went to court, asking for a legal separation, ownership of the ranch home and the land surrounding it, and a one-half interest in the ranch land, cattle, cattle brand, and other assets. *The court granted her only $200 a month maintenance.* Everything else—land, cattle, assets, family home—went to Jim. The judge justified his ruling this way: "The land was held in the name of Mr. Murdoch at all times. The cattle and equipment were also held in his name; no declaration of partnership was ever filed . . . and I, therefore, do not form the conclusion that the Murdochs were partners, or that a relationship existed that would give Mrs. Murdoch the right to claim as a joint owner in equity in any of the farm assets."

Irene Murdoch had worked for 25 years to make her husband a rich man. Her reward was a severe beating and the promise of $200 a month. Similar cases are to be found in almost all common-law states.

[a]Dower rights are the historical rights of the surviving wife to a life estate in one third of the land her husband owned at any time during the marriage. In Wisconsin and some other states dower rights have been abolished and replaced by a statutory "forced share."

Source: Based on *Real Women, Real Lives* (Madison: Wisconsin Governor's Commission on the Status of Women, 1978), pp. 44–45.

---

spectives united in one contradiction. Under current conditions, to eliminate the "hidden" exploitation within farm families would be to destroy family farms. After all, male farmers are not intrinsically evil, but their own exploitation and their inability to compensate their families for labor contributions make the use of "free" labor a necessary and normal way of operating family farms under agrarian capitalism.

The economic efficiency of family farmers has long been recognized by agricultural economists, but this conventional analysis ignores the labor contributions of female farmers and farm children as well as the unusually low wage incomes of male farmers, upon whom the farm income data are based. If the labor value of all farm family members were included, the efficiency of family farms would be seen instead as excessive exploitation. Much of the staying power of family farmers (in their conventional form) is not their productive efficiency, but their "willingness" to operate at a deficit for long periods of time—that is, to accept a low standard of living and to suffer extreme forms of exploitation. Indeed, what remains of family farming is assured only by government payments through the various farm programs, off-farm employment, and financial indebtedness.

Large-scale and small-scale farmers stand in opposition to each other. The major role of large-scale farmers, with their reliance on hired labor and capital-intensive production, is as capitalists. On the other hand, small-scale farmers function predominantly as workers. Family farmers who sign production contracts with processing plants lose their management role and resemble industrial workers even more than other family farmers. Although tenants make annual decisions about farm production, their access to land is restricted to cash leases with landlords. Like tenants, sharecroppers have limited access to land and hand over most of their decision-making powers to their landlords. Finally, hired farm workers perform only one role, that of workers, which complements the role of agricultural managers. Both are hired by capitalists (corporate or private) to operate industrial and large-scale farms.

The multiple roles that many rural groups play result in dialectical social relations that result in conflict and struggle. Table 14.4 illustrates the kinds of dominant/dependent relations one group holds relative to other groups. National farm input and processing firms dominate farmers but are dependent on multinational corporations. Similarly, dominant and dependent class relations occur for all other rural groups until, ultimately, migratory farm workers

TABLE 14.4
Examples of Class Conflicts in U.S. Agriculture

	Dominant	Dependent	
	capitalists	workers	
	multinationals	farm input & output firms	
	farm input & output firms	all farmers	
	corporations in farming	family farmers	
	family farmers	hired farm workers	
	farm husbands	farm wives	
		farm children	
		occasionally relatives	
	female farmers	farm children	⎤
whites ⎡	permanent farm workers	seasonal farm workers	⎤ racial
	seasonal, local farm workers	migratory farm workers	⎦ minorities
⎣	U.S. migratory farm workers	undocumented farm workers ⎦	

with U.S. citizenship have rights and power over the lowest group, illegal aliens. Each group tries to maintain its present power with respect to a dependent group and at the same time tries to increase its power with respect to a dominant group.

Family farmers in the United States have nearly ceased to exist as independent entrepreneurs. They are increasingly and rapidly being turned into outdoor production workers, although most continue to believe that they are independent businesspeople. Farmers are particularly susceptible to this contradiction because of their employment characteristics and their own conflicting roles. Technically, farming still occurs under "free-market" conditions because the majority of farm operators are self-employed. In 1969 only 9 percent of the *total* U.S. labor force was so employed—most of whom were farmers. Hence, agriculture is the only major economic sector that at least *resembles* the ideal capitalist society. This statistic allows farmers, politicians, and the general public to think that all is well on the farm. The illusion that family farms predominate in U.S. agriculture and that competition is characteristic of this industry obscures the deep-seated corporate and governmental forces that have concentrated the bulk of agricultural production in the hands of large-scale farmers while family farmers are destroying themselves in fierce competition.

In addition to the dialectics of dependence and independence among classes and subclasses, there also exists the dialectics of similarity and dissimilarity among workers. The interests of agribusiness and capitalism in general are best served by a homogeneous work force that can be easily interchanged and manipulated. A natural result of capitalism, then, is the homogenization of workers as well as production processes. As family farmers are forced more and more into the proletariat, they take on more characteristics of hired agricultural labor and even industrial labor. Of course, the threat of such homogenization is that workers then come to recognize their similarities and are strengthened as a class for a struggle against capital. Thus, "the only way for capital to achieve its need for the controllable homogeneity of abstract labor is, paradoxically, through the imposition of heterogeneity, through the division of workers. It is only by dividing and pitting one group of workers against another that capital can prevent their dangerous unity and keep the class weak enough to be controlled."[18] The myth of the family farm provides the necessary division of workers by making farmers feel different from other workers while camouflaging their structural similarities. In a Marxian analysis, "Family farmers form a class the way a sack full of potatoes forms a class. They all have the same characteristics and are a class-in-itself, but because they fail to act together politically they do not form a class-for-itself."[19] By maintaining a confusion between the real and ideal roles of farmers, the dialectics of homogeneity/heterogeneity benefits agribusiness and destroys the family farm.

Conclusion

The first American Revolution resulted in political freedoms; the second will be for economic freedoms. Democracy based on individual political rights is meaningless for people without economic rights, and civil rights are granted in liberal democracies only when they do not threaten the basic economic system. The next revolution in the U.S. will require the expansion of political democracy to include economic democracy; only then can agrarian capitalism give way to agrarian democracy. Because of the increasingly contradictory forces U.S. family farmers face in the 1980s, they need to work toward this democratic and economic revolution.

Once family farmers realize that they are a part of the working class, they can work with other groups sharing the same class basis. Structurally, family farmers have more in common with other workers than with large-scale producers and agribusiness firms. Their isolated outdoor work, rural location, and seasonal rhythms of agriculture obscure the real class interests of family farmers that are shared with nonfarm workers. To achieve political and economic equality, family farmers cannot identify with their oppressors, with rural institutions, or with groups not having their best interests at heart and in mind. As long as farmers have hope in the conventional politics of the two-party system, they achieve at best liberal reforms, rather than the fundamental solutions necessary for their well-being and survival. Family farmers today need extraordinary changes, not ordinary reforms. If farm movements are to have a significant impact, as the Populists did, the underlying values of the larger society must be challenged. A class analysis and a working-class perspective are critical for a progressive farm movement to emerge and build alliances with like-minded urban groups—and to ultimately achieve democratic socialism and true agrarian democracy.

Dialectically, family farmers can turn the very instruments of their oppression and extinction into their liberation. The *myth* of the family farm can be destroyed through self-education and group action. Through the democratic process, progressive farmer-labor alliances can turn the ideals of the family farm into reality and turn agribusiness into a vanishing species. In the 1870s industrial workers lacked the organizational strength to join the insurgent politics of the Populists. A hundred years later, many national labor unions or their locals are sufficiently organized and radical to form alliances with farmers, but farmers are insufficiently organized and progressive to reciprocate.[20] Today's generation of family farmers may have the last opportunity to achieve a farmer-labor coalition.[21]

In the 1980s farmers and other workers need to rediscover their radical roots and finally forge a unity to establish a truly democratic political and economic society. The "movement culture," in Lawrence Goodwyn's words, exists today among progressive women, racial minorities, the poor, and environmental, religious, and labor groups.[22] These groups and family farmers share common ideals and class relations—the basis for political solidarity. The Minnesota Democratic Farmer-Labor Party, which was an offshoot of the Non-Partisan League in the 1930s, is the best-known example of such an alliance today. In 1977, the progressive spirit of the farmer-labor coalitions of the 1920s and 1930s was revived by the formation of the Farmer-Labor Association (FLA). The FLA believes that corporate power rules business and government and "only the people who make the machines, teach the children, till the land, raise the families" can change the United States for the better.[23]

Working together, consumers and farmers can have both lower food costs and higher agricultural prices if they share even a fraction of the immense profits made by the food industry. Such sharing would lead to greater economic justice and greater equality of income distribution. Alliances between all working-class people would transform them from a class-in-itself, exchanging their labor power for income, to a class-for-itself, discovering their unity and strengthened through struggle against capitalism.

To shatter the myth of the family farm and to build within-class alliances will be difficult. Most family farmers are already working for progressive causes when they believe in democracy, equality, and justice—concepts that are certainly not self-evident and require long-term commitments to human well-being to achieve. By trying to change the status quo to achieve these ideals in agriculture, we will gain a greater understanding of how U.S. agriculture really works. This understanding will be the basis for a new populism. May we seize the opportunity.

Appendix:
Organizations Working for
Progressive Rural Change

American Agri-Women (AAW)
6690 Walker Avenue, N.W.
Grand Rapids, Michigan 49504
Phone: (616) 784-2821
Sharon Steffens, national coordinator

Women's agricultural organizations and individuals united together to communicate with one another and with other consumers to promote agriculture for the benefit of the American people and the world. *Rural focus:* supports farmer cooperatives, investigates local land-use proposals, and supports estate-tax reform.

American Country Life Association
2118 South Summit Avenue
Sioux Falls, South Dakota 57105
Phone: (605) 336-5236
Osgood Magnuson, president

A voluntary organization of people in various private and public walks of life interested in current issues and ideas affecting the economics, well-being, and quality of life in nonmetropolitan areas of the country. Its main activity is convening annual forums that concentrate on some issue or theme of pertinent concern to the long-range welfare and progress of rural areas. *Rural focus:* various aspects of rural development.

American Friends Service Committee (AFSC)
1501 Cherry Street
Philadelphia, Pennsylvania 19102
Phone: (215) 241-7000
Louis W. Schneider, executive secretary

Regional offices in ten U.S. cities. Related to the Religious Society of Friends, the AFSC attempts to relieve human suffering and to find new approaches to world peace and nonviolent social change. Activities include

Largely based on *A Directory of Rural Organizations* (Washington, D.C.: National Rural Center, 1971).

working for basic human rights, peace education, providing technical and medical assistance abroad, and providing opportunities for reconciliation and peace building both at home and abroad. *Rural focus:* various programs related to fighting poverty and injustice including work with Native Americans, land- and food-control issues in California, rural housing, hunger, and human rights.

American Institute of Cooperation
Suite 504, 1129 20th Street, N.W.
Washington, D.C. 20036
Phone: (202) 296-6825
Owen K. Hallberg, president

A national educational organization supported by agricultural cooperatives. Educational programs and materials are designed to help acquaint the public with the role of cooperatives in the American competitive enterprise system. Educational and training services are made available through a national agribusiness institute, several regional institutes, publications, and various consulting activities by the staff. *Rural focus:* rural and agricultural cooperatives.

American Medical Association–Department of Rural and Community Health
535 North Dearborn Street
Chicago, Illinois 60610
Phone: (312) 751-6422
Richard E. Palmer, president   James H. Sammons, exec. vice-president

Aims to improve access to and quality of health and medical care for rural areas and to provide positive response to national health planning issues and home health care. *Rural focus:* delivery of rural health services, community organization for rural health services, and health education for rural people for personal health improvement.

Association on American Indian Affairs, Inc.
432 Park Avenue South
New York, New York 10016
Phone: (212) 689-8720
Alfonso Ortiz, president   William Byler, executive director

Promotes the welfare of American Indians in the United States by creating an enlightened public opinion; by assisting and protecting them against violations of their constitutional rights; by aiding in the improvement of health and educational conditions; and by preserving their arts and crafts. Aids tribes in mobilizing federal, state, and private resources for a coordinated attack on the problems of poverty and injustice. *Rural focus:* providing legal, technical, and financial assistance programs to Indian communities.

Association for Rural Mental Health
610 Langdon Street
Madison, Wisconsin 53703
Phone: (608) 263-2088
Cecil M. Hudson, president

Represents the interests of rural mental health programs in dealing with federal agencies and national mental health associations. Sponsors continuing education activities for staff members in rural mental health centers and establishes study committees.

The Box Project, Inc.
P.O. Box 435
Plainville, Connecticut 06062
Phone: (203) 747-8182
Sybille Brewer, president   Richard Barnett, executive director

Assists rural poverty families with their basic needs and exposes the problems of poverty to middle-income American families in the hope that this will provide a stop-gap type of help to families living in poverty "until such time as the federal government makes the decision to eliminate poverty in the United States." *Rural focus:* arranges for sponsor families to assist needy families who are referred by field workers, community centers, and personal appeals in such areas as housing, education, welfare benefits, food, and clothing. Needy families are located in the rural South and on Native American reservations in the upper Midwest.

Bread for the World
207 East 16th Street
New York, New York 10003
Phone: (212) 751-3925
Eugene Carson Blake, president   Arthur Simon, executive director

A Christian citizens' lobby movement concerned with issues of hunger, such as assistance to poor countries, trade, military spending, and investment and employment. Members are urged to work with others in local hunger action groups and to contact members of Congress and other government leaders regarding U.S. policy matters that affect hungry people. *Rural focus:* supporting a fair return to U.S. farmers for their production, curbs against windfall profits, special measures to assist family farmers, just wages for farm workers, and appropriate control of multinational corporations with particular attention to agribusiness.

Center for Community Change
1000 Wisconsin Avenue
Washington, D.C. 20007
Phone: (202) 338-6310
Pablo Eisenberg, president

Provides technical assistance and other support to low-income and minority community organizations throughout the country. Support includes help in program planning and implementation, large-scale physical development projects, resource development, internal administration, training and workshops, and general information. Does not provide financial assistance to local groups. *Rural focus:* has engaged in a variety of technical assistance, intervention, public information, and advocacy activities with groups of farm workers, small farmers, Appalachians, and other rural minorities. Special projects have included research on rural education, the Agribusiness Accountability Project, and the Youth Project, which has studied the effects of large-scale strip mining on the northern plains.

Center for New Corporate Priorities
Suite 202, 1516 Westwood Boulevard
Los Angeles, California 90024
Phone: (213) 475-5856
Jim Lowery, director

Seeks economic reform, focusing on banks and credit policy. Researches the environmental impact of bank lending decisions. Its National Task Force on Credit Policy has sought new credit policies to better serve minorities, women, small farmers, and urban dwellers. *Rural focus:* researches agribusiness financing.

Center for Rural Affairs
P.O. Box 405
Walthill, Nebraska 68067
Phone: (402) 846-5428
Don Ralston and Marty Shange, codirectors

Promotes rural development within Nebraska and throughout the United States by providing information on the trends and implications of changes in government, agriculture, and private industry; by providing opportunities to discuss these trends and their implications; and by conducting research and publishing reports on developments in rural Nebraska. *Rural focus:* small farm energy, small farm advocacy, and rural banks.

Center for the Study of Local Government
22 Fifth Avenue South
St. Cloud, Minnesota 46301
Phone: (612) 253-7110
L. Dennis Kleinsasser, director

Activities on individual, group, organization, or community levels. Provides packaged instructional materials concerning land-use planning, personnel management, supervisory training, and school governance. Organizational consultation includes organizational development, team building, process consultation, and conference design expertise. Affiliated with St. John's

University. *Rural focus:* working to make rural local government officials more knowledgeable and effective in their governmental roles.

Congressional Rural Caucus (CRC)
309 House Annex Building
Washington, D.C. 20515
Phone: (202) 225-5080
John B. Breckenridge, chairman   Frank G. Tsutras, director

A bipartisan group of U.S. representatives dedicated to the orderly growth and development of rural America. Seeks to combine maximum federal and nongovernmental resources available to rural communities throughout the nation. *Rural focus:* rural education, health, public works, transportation, sewage and water treatment, housing, financial resources and credit requirements, implementation of the Rural Development Act of 1972, and job opportunities.

Cooperative League of the U.S.A.
1828 L Street, N.W.
Washington, D.C. 20036
Phone: (202) 672-0550
Stanley D. Dreyer, president   Shelby E. Southard, rural affairs

National federation of customer-owner businesses including farm marketing and supply, rural electric, farm credit, rural telephone, credit unions, mutual insurance, housing, and health. Services provided for member organizations include advocacy on legislation, liaison with government, education, organization, and training. *Rural focus:* adequate funding for rural development focusing on preservation of the family farm, increased direct marketing from farm to urban consumer, better farmer-consumer relations, adequate rural medical care and transportation, and preservation of the rights of farmers to organize under the Copper-Volstead Act.

Emergency Land Fund (ELF)
836 Beecher Street, S.W.
Atlanta, Georgia 30310
Phone: (404) 758-5506
Robert S. Browne, president   Joseph F. Brooks, executive director

Washington Office:
Suite 500, 1346 Connecticut Avenue, N.W.
Washington, D.C. 20036
Phone: (202) 293-1649, Gloria Cousar

Believes that broad-based landownership by blacks in the rural South is in the best interest of the larger black community as well as the nation, and, further, that through landownership blacks can maintain the option of remaining in rural areas. Seeks to address the decline of black land loss, with the hope of ultimately reversing the land-loss trend and increasing black land holdings. Activities include educational and instructional outreach, land-use demonstration, and applied research. *Rural focus:* concerned about discrimination against black landowners in the rural South; seeks to redress the absence or underextension of capital, markets, and other technical development resources for black landowners or would-be landowners. Provides legal, financial, management, and technical assistance for black landowners or would-be landowners in Mississippi, Alabama, Georgia, and South Carolina, and provides extended services to other Southern states.

Eric Clearinghouse on Rural Education and Small Schools
Box 3AP, New Mexico State University
Las Cruces, New Mexico 88003
Phone: (505) 646-2623
Everett D. Edington, director

A national information system operated by the National Institute of Education. Provides ready access to descriptions of exemplary programs, research and development efforts, and related information that can be used in developing more effective educational programs. Responsible for acquiring, indexing, abstracting, and disseminating information related to all aspects of education of American Indians, Mexican Americans, migrants, as well as outdoor education and education in small schools and rural areas. Directs workshops and provides consultation services. *Rural focus:* education.

Exploratory Project for Economic Alternatives (EPEA)
Room 515, 2000 P Street, N.W.
Washington, D.C. 20036
Phone: (202) 833-3990
Gar Alperovitz and Jeff Faux, codirectors

Established by a consortium of twenty-five foundations to help determine how the United States might reshape its economic system over the coming decades around principles of democracy, cooperation, fairness, decentralization, and the efficient and conserving use of resources. Seeks to broaden citizen debate through the development of specific proposals in such areas as: national food policy, cooperatives, public trusts for conservation, community-based planning, housing, resource allocation, and public enterprise. *Rural focus:* EPEA's food project has recommended substantial restructuring of the U.S. food system toward low consumer prices, stable family farms, and environmentally sound agriculture. EPEA's food project is developing an educational campaign around a comprehensive national food policy for urban and rural America by working with farm, church, environmental, and consumer groups.

Family Farm Education Project (NFFEP)
Room 624, 815 15th Street, N.W.
Washington, D.C. 20005
Phone: (202) 638-6848

Educates the general public about the need for an agriculture system and government policies that support and encourage small and moderate-sized family farms. *Rural focus:* clearinghouse for congressional information about family farming.

Mexican American Legal Defense and Educational Fund, Inc. (MALDEF)
145 Ninth Street
San Francisco, California 94103
Phone: (425) 864-6000
Vilma S. Martinez, president and general counsel

Washington Office:
1028 Connecticut Avenue, N.W.
Washington, D.C. 20036
Phone: (202) 659-5166
Al I. Perez, associate counsel

Advances the civil rights of Chicanos and Latinos through legal action and legal education. Litigates class-action suits that strive to secure equal employment and educational opportunity and equal political access for Hispanic Americans. Available services include: loan forgiveness/scholarship funds for Chicano law students, community education and activation, and legal and sociological research into issues affecting the Chicano community. *Rural focus:* has handled land and water rights cases for Chicanos in rural areas of Texas and New Mexico. Political access litigation and equal educational opportunity litigation has focused on the needs of rural Chicanos, particularly in South Texas.

Migrant Legal Action Program, Inc.
1910 K Street, N.W.
Washington, D.C. 20006
Phone: (202) 785-2475
Raphael Gomez, executive director

Seeks to defend the rights of economically deprived agricultural workers. Engages in legislative monitoring and federal agency advocacy and provides neighborhood legal services for people unable to secure legal counsel, particularly farm workers. *Rural focus:* national advocacy centers on issues of worker's compensation, minimum wage, occupational safety and health standards, housing, food stamps, and fair labor standards.

National Association for the Advancement of Colored People (NAACP)
1790 Broadway
New York, New York 10019
Phone: (212) 245-2100
Benjamin Hooks, executive director

Seeks to insure the political, educational, social, and economic equality of all citizens; to achieve equality of rights and eliminate race prejudice; and to remove all barriers of racial discrimination through democratic processes. Every department at national headquarters provides guidance and expertise to assist regional offices, state conferences, and individual branches (rural and urban) with their programs by: organizing and participating in workshops as well as arranging for outside specialists to participate; sending information and action guides to branches; providing advice on current problems; and advising and monitoring government activity at all levels. *Rural focus:* the Emergency Relief Fund is aimed chiefly at the rural population in alleviating hunger in critical areas of the South. Voter education and registration programs also apply to rural areas.

National Association of Farmworker Organizations
Suite 1145, 1328 E Street, N.W.
Washington, D.C. 20004
Phone: (202) 347-2407
Humberto Fuentes, president   Tom Jones, national representative

Advocates coherent federal policies and services to farm workers and rural poor. Acts as a clearinghouse for farm-worker programs and as a catalyst for farm workers and their organizations. Provides training and technical assistance. Monitors and analyzes legislation, federal policies, and regulations and their impact on farm workers. *Rural focus:* employment, housing, health, economic development, child development, and other programs or policies that affect farm workers.

National Catholic Rural Life Conference
3801 Grand Avenue
Des Moines, Iowa 50312
Phone: (515) 274-1581
John J. McRaith, executive director   Stephan E. Bossi, research director

Promotes "the responsible development and just distribution of earth's resources and the fruits thereof, rural community and economic development, and improved Church ministry in the rural community." Provides educational materials, workshops and training sessions in natural resources and rural issues, a speakers bureau, advocacy with government agencies and officials, and research and resource assistance. *Rural focus:* (1) natural resource policy—land use, energy development, and soil conservation; (2) food policy—farm programs, vertical integration, and corporate agriculture; (3) rural development—housing, health care, and employment; (4) rural ministry—training of ministers and encouraging local parish activities in rural areas.

National Child Labor Committee
145 East 32nd Street
New York, New York 10016
Phone: (212) 683-4545
Ronald H. Brown, president   Jeffrey Newman, executive director

Seeks to promote the rights and dignity of children and youth at work. *Rural focus:* serves as a clearinghouse and advocate for the education of migrant children. Maintains a small emergency scholarship program for migrant students in college. Seeks protection of young children from work abuse, particularly in agriculture. Serves as an advocate for unemployed and under-employed youth.

National Community Land Trust Center
Suite 316, 639 Massachusetts Avenue
Cambridge, Massachusetts 02139
Phone: (617) 661-4661
Robert Swann, director

Affiliated with the International Independence Institute. Serves the community land-trust movement as a clearinghouse and a research/resource center and conducts an aggressive outreach program. Provides technical information, news, and information about community land trusts. Maintains a land-exchange file of community land trusts having land available and of individuals seeking access to land. *Rural focus:* promotes community land trusts that are legal entities and that acquire land for responsible community uses.

National Congress of American Indians/NCAI Fund
Suite 700, 1430 K Street, N.W.
Washington, D.C. 20005
Phone: (202) 347-9520
Charles Trimble, executive director

Serves as an information clearinghouse and a legislative advocate for American Indian tribes. Seeks to protect, conserve, and develop Indian land, mineral, timber, and human resources and to improve the health, education, and economic condition of American Indians. *Rural focus:* protection of reservation lands and resources and the improvement of socioeconomic conditions of rural Native Americans.

National Council of Farmer Cooperatives
1129 20th Street, N.W.
Washington, D.C. 20036
Phone: (202) 659-1525
Kenneth D. Naden, president   Paul S. Weller, vice-president

A nationwide association of cooperative businesses owned and controlled by farmers. Represents farmer cooperatives on national and international matters and works closely with Congress, the executive branch, and federal regulatory agencies. *Rural focus:* rural development, taxation, marketing, trade, farm finance, supply, environmental concerns, grain marketing, transportation, and farm labor.

National Council of La Raza
Suite 210, 1725 I Street, N.W.
Washington, D.C. 20005
Phone: (202) 659-1251
Raul Yzaguirre, national director   Eduardo Terrones, deputy director

Serves as a nationally based research and advocacy organization for Spanish-speaking community groups throughout the country, voicing their concerns to the agencies and institutions charged with responding to their needs and problems. *Rural focus:* provides technical assistance to Spanish-speaking individuals and groups in areas such as employment, housing, education, health, and economic and business development.

National Farmers Organization
720 Davis Avenue
Corning, Iowa 50841
Phone: (515) 322-3131
Oren Lee Staley, president

Legislative Office:
475 L'Enfant Plaza, S.W.
Washington, D.C. 20024
Phone: (202) 484-7075
Charles L. Frazier, director   Ann Bornstein, legislative assistant

A nonpartisan organization of farmers bargaining collectively for the sale of farm commodities. Seeks to assure the survival of the family farm in America. *Rural focus:* includes support of the Family Farm Antitrust Act of 1975, farm programs, programs for conservation of soil and water, rural electrification and telephone, and farm bargaining legislation.

National Farmers Union
P.O. Box 39251
12025 E. 45th Avenue
Denver, Colorado 80239
Phone: (303) 371-1760
Tony T. Dechant, president

Washington Office:
1012 14th Street, N.W.
Washington, D.C. 20036
Phone: (202) 628-9774
Robert G. Lewis, national secretary

A family farm organization that carries on a broad range of educational and legislative activities. Promotes programs to assure the continuation and strengthening of the individual family-type farm as the primary system of agriculture in the United States. Develops and promotes programs to strengthen family farming and rural life and espouses new national food, farm energy, and land policies. *Rural focus:* supports programs to promote the general welfare of rural America including conservation, rural development, health, full employment, electrification, and the development of farmer cooperatives.

National Land for People
Room 11, 1759 Fulton
Fresno, California 93721
Phone: (209) 233-4727
George Ballis, president   Geneva Gillard, chief staff person

Does research, litigation, and public education on large issues of peoples' legal land and water rights. Focus is on enforcement of the 160-acre limitation (Reclamation Law), "which if enforced, would break-up corporate holdings throughout the 11 Western States and primarily in the San Joaquin Valley." Also interested in showing possibilities for application of appropriate technology and alternative energy to the small farmers who will populate the land, as well as encouraging land trusts and cooperative ventures. Conducts research on land-ownership patterns and shares research skills with other organizations. Helps others start food-buying clubs. Media presentations are available to educate people about the importance of retaining healthy small towns by encouraging small farms. *Rural focus:* believes that water-land policies have urban as well as rural significance and local, national, and international ramifications.

National Rural Center (NRC)
1828 L Street, N.W.
Washington, D.C. 20036
Phone: (202) 331-0258
John M. Cornman, president

An independent, nonprofit organization created to develop policy alternatives and to provide information that can help rural people achieve full potential. Develops policy alternatives by conducting demonstration programs, using the results of existing research, evaluating programs, and, where needed, conducting basic research. Information clearinghouse helps state and local officials, community organizations, and individuals with rural concerns find information about and gain access to sources of federal and private funds and expertise. Maintains a library of research and data on rural affairs, including legislative and executive documents, and offices in Atlanta, Georgia, and Austin, Texas. *Rural focus:* all aspects of rural development and revitalization including health, transportation, communications, employment, migration trends, legal services, education, small-farms policy, and community economic development.

National Sharecroppers Fund/Rural Advancement Fund
2128 Commonwealth Avenue
Charlotte, North Carolina 28205
Phone: (704) 334-3051
Benjamin E. Mays, chairman   James M. Pierce, executive director

The National Sharecroppers Fund seeks the enactment of national legislation that promotes and protects the interests of rural poor people. The central mission of the Rural Advancement Fund is the improvement of the social and economic conditions of America's rural poor through agricultural skill training and the advancement of community-based cooperative organizations. Conducts educational programs to acquaint the public with problems of rural America and their relationship to urban problems. *Rural focus:* major activity is rural training at the Frank P. Graham Center near Wadesboro, North Carolina.

Network
224 D Street, S.E.
Washington, D.C. 20003
Phone: (202) 544-1371
Carol Coston, executive director

A Catholic lobby group seeking national legislation that is responsive to the needs of all people, especially the poor, the hungry, the jobless, and the imprisoned. Conducts an annual referendum of its membership to determine issue priorities. Current issues include: equitable food planning and food programs at home and abroad, national health insurance, full employment and welfare reform, alternatives to incarceration, and decreased militarization. *Rural focus:* preservation of family farms and support for land reform.

Public Citizen's Tax Reform Research Group
133 C Street, S.E.
Washington, D.C. 20003
Phone: (202) 544-1710
Ralph Nader, president, Public Citizen, Inc.; Robert M. Brandon, director

Seeks to reform the tax systems in America to make them equitable, simple, and efficient, and to inform the public of how the tax systems work. Provides information on tax matters and (where resources permit) assistance to local groups interested in tax reform. Lobbies for beneficial tax reform in Congress. *Rural focus:* as part of basic tax reform, seeks to eliminate farming tax shelters and to make property taxation fairer. Also, seeks a fair, somewhat redistributive estate-tax system that recognizes the special problems of farmers.

Rural America
1346 Connecticut Avenue, N.W.
Washington, D.C. 20036
Phone: (202) 659-2800
Clay Cochran, executive director   Roger Blobaum, chairperson

A nonprofit membership organization established to meet the need for continuing national advocacy on behalf of rural people. The main goals are to assure rural people equity in the formulation and implementation of federal policy and to serve as a national clearinghouse for information and services to rural individuals and groups. Carries out a program of policy-oriented research, technical assistance and training, legislative information, and public education and advocacy. Its rural housing program provides technical and financial assistance to farm-worker and other local housing groups. Has established rural housing and health councils and seeks to encourage formation of other special-interest coalitions.

United Farm Workers of America, AFL-CIO
P.O. Box 62
Keene, California 93531
Phone: (805) 822-5571
Cesar Chavez, president

Organizes farm workers; dedicated to obtaining better wages, improved working conditions, job security, and protective contracts.

United Methodist Church Women's Division "Family Farm Project"
Box 316
LaFarge, Wisconsin 54639
Phone: (608) 625-4505
Jane Johnson, project organizer

Methodist women organized in thirty-seven of the fifty-five Methodist conferences to promote the survival of the family farm and the values it contributes to society. Encourages stewardship of land and resources for present and future generations. *Rural focus:* participants are involved in land-tenure studies, monitoring land-use and farm-labor legislation, soil surveys to estimate productivity of soil in case of future fertilizer shortages, land-grant college reviews, estate tax reform, and OSHA regulations.

Notes

Chapter 1   Introduction: The Myth of the Family Farm

1. *1978 Handbook of Agricultural Charts*, USDA (Washington, D.C.: Government Printing Office, 1978).

2. Irving Howe, "The Right Menace," *New Republic*, September 9, 1978, p. 16.

3. David H. Ciscel, "Financial Dogs," *New Republic*, June 2, 1979, p. 8.

4. "Fred Harris Runs It Down," *People and Land*, Summer 1973, p. 10.

5. Herbert I. Schiller, *The Mind Managers* (Boston: Beacon Press, 1973), p. 24.

6. Ibid., p. 1.

7. The attachment to a conservative ideology is more pronounced among corn farmers in the Midwest than among wheat farmers in the Great Plains. See Jeffery M. Paige, *Agrarian Revolution* (New York: Free Press, 1975), pp. 36-37.

8. John Fraser Hart, "The Middle West," in John Fraser Hart (ed.), *Regions of the United States* (New York: Harper and Row, 1972), p. 272.

9. Paige, *Agrarian Revolution*, p. 32.

10. See Richard J. Barnet, *The Crisis of the Corporation* (Washington, D.C.: Transnational Institute and Institute for Policy Studies, 1975), pp. 32-33.

Chapter 2   What Is a Family Farm?
              Farm Definitions and Classifications

1. B. Delworth Gardner and Rulon Pope, "How is Scale and Structure Determined in Agriculture?" *American Journal of Agricultural Economics* 60 (May 1978), p. 300.

2. Radoje Nikolitch, "Family-Operated Farms: Their Compatibility with Technological Advance," *American Journal of Agricultural Economics* 51 (1969), p. 531; italics added. No such detailed data has been made available since 1969.

3. Radoje Nikolitch, *Family-Size Farms in U.S. Agriculture*, USDA, Economic Research Service Report No. 499 (Washington, D.C.: Government Printing Office, 1972), p. 3.

4. Ibid.

5. Ibid.

6. Ibid.

7. Nikolitch, "Family-Operated Farms," p. 533.

8. Ibid., p. 536.

9. U.S., Congress, House, *Family Farm Act of 1972, Hearings* before the Anti-Trust Subcommittee of the Committee on the Judiciary, 94th Cong., 1st sess., 1975, p. 20 (hereinafter *Family Farm Act of 1972*).

10. Nikolitch, "Family-Operated Farms, " p. 536.

11. Ibid.

12. Richard D. Rodefeld, "A Reassessment of the Status and Trends in 'Family' and 'Corporate' Farms in U.S. Society," *Congressional Record*, 93d Cong., 1st sess. (May 31, 1973), p. S10056.

13. Ibid.

14. Ibid.

15. Ibid., p. S10057.

16. Nikolitch, "Family-Operated Farms," p. 531.

17. Rodefeld, "A Reassessment of Status and Trends," p. S10057.

18. Phil Margolis, "AG Census Misleading," *Rural America* 2, no. 6 (April 1977), pp. 5 and 7. Even more drastic measures have been proposed by the Bureau of the Census. J. Thomas Breen, chief of the census bureau's agricultural division, would like to eliminate the demographic aspects of agriculture and concentrate only on the business of farming, which is agribusiness. The bureau would also like to eliminate farms by defining them by a minimum level of sales. Various levels have been suggested. If the minimum value of sales is set at $2,500, that would eliminate 42 percent of all farms in the U.S. At $5,000, 56 percent of all farms would be eliminated and at $10,000, 72 percent of all farms would disappear! See Jim Hightower and Susan Sechler, *Counting Out Farmers* (Washington, D.C.: Agribusiness Accountability Project, February 1973).

19. Rodefeld, "A Reassessment of Status and Trends," p. S10057.

20. Ibid., p. S10058.

21. Nikolitch, *Family-Size Farms*, p. 17 and table 10.

22. For a detailed discussion of this classification, see Richard D. Rodefeld, "The Changing Organizational Structure of Farming and the Implications for Farm Work Force, Individuals, Families and Communities" (Ph.D. diss., University of Wisconsin, 1974).

23. Elsewhere Rodefeld considers gross farm sales and acreage operated. See Richard D. Rodefeld, "Trends in U.S. Farm Organizational Structure and Type," in U.S., Congress, Senate, *Priorities in Agricultural Research of the U.S. Department of Agriculture, Hearings* before the Subcommittee on Administrative Practice and Procedure of the Committee on the Judiciary, 95th Cong., 2d sess., 1978, part 1, pp. 374–375.

24. The Bureau of the Census is planning a computer user file tape that would allow users, for a fee, to create their own cross tabulations of any set of variables found in the Census of Agriculture. This procedure will eliminate the single-variable definition of the family farm, which this and previous work have had to rely on.

25. Ed Schaefer, *Changing Character and Structure of American Agriculture: An Overview* (Washington, D.C.: General Accounting Office, 1978).

26. Nikolitch, "Family-Operated Farms," p. 532.

27. Lawrence Goodwyn, "The Neoconservatives: The Men Who are Changing America's Politics" (book review), *New Republic*, September 1 and 8, 1979, p. 29.

28. Gardner and Pope, "How is Scale and Structure Determined in Agriculture?" p. 301.

29. Ibid., p. 302.

30. Fred Schmidt, *Issues on Agricultural Research* (Washington, D.C.: Rural America, 1977); and *Family Farm Act of 1972*.

31. Catherine Lerza, "Farmers on the March," *The Progressive*, June 1978, pp. 22-25.

32. Marshall Harris, *Entrepreneurship in Agriculture*, Monograph 12 (Iowa City: University of Iowa, College of Law, 1974), p. 11.

Chapter 3   U.S. Land-Granting Policies

1. Philip M. Raup, "Societal Goals in Farm Size," in A. Gordon Ball and Earl O. Heady (eds.), *Size, Structure, and Future of Farms* (Ames: Iowa State University Press, 1972), pp. 3-18.

2. Paul Wallace Gates, "An Overview of American Land Policy," in Vivian Wiser (ed.), *Two Centuries of American Agriculture* (Washington, D.C.: Agricultural History Society, 1976), p. 217.

3. Benjamin Horace Hibbard, *A History of the Public Land Policies* (Madison: University of Wisconsin Press, 1965), p. 219.

4. Ibid., pp. 30-31.

5. Ibid., pp. 45-49.

6. Sheldon L. Greene, "Promised Land: A Contemporary Critique of Distribution of Public Land by the United States," *Ecology Law Quarterly* 5, no. 4 (1976), p. 710.

7. Ibid., p. 711.

8. Hibbard, *History of the Public Land Policies*, p. 24.

9. For a thorough discussion of early land concentration in California, see U.S., Congress, Senate, *Farmworkers in Rural America, 1971-1972, Hearings* before the Subcommittee on Migratory Labor of the Committee on Labor and Public Welfare, 92d Cong., 2d sess., 1972, appendix, part 5B, pp. 3941-3947.

10. Paul Wallace Gates (ed.), *California Ranchos and Farms, 1846-1862* (Madison: State Historical Society of Wisconsin, 1967), pp. 7-8.

11. Greene, "Promised Land," p. 712.

12. Gates, *California Ranchos and Farms*, p. 16.

13. Ibid., p. 15.

14. Cited in Paul S. Taylor, "Mexican Migration and the 160-Acre Water Limitation," *California Law Review* 63, no. 3 (1975), p. 739.

15. Cited in Mary W. M. Hargreaves, "Women in the Agricultural Settlement of the Northern Plains," in Vivian Wiser (ed.), *Two Centuries of American*

Agriculture (Washington, D.C.: Agricultural History Society, 1976), p. 186.

16. Quoted in Hibbard, *History of the Public Land Policies*, pp. 369-370.

17. Quoted in ibid., p. 371.

18. Ibid., p. 380.

19. Greene, "Promised Land," p. 713.

20. Hibbard, *History of the Public Land Policies*, p. 386.

21. Quoted in ibid., p. 389.

22. Paul Wallace Gates, "The Homestead Law in an Incongruous Land System," in Vernon Carstensen (ed.), *The Public Lands* (Madison: University of Wisconsin Press, 1968), p. 318.

23. Ibid., p. 319.

24. Ibid., p. 325.

25. Ibid., p. 328.

26. Ibid., p. 329.

27. David Maldwyn Ellis, "Comment on 'The Railroad Land Grant Legend' in American History Texts," in Vernon Carstensen (ed.), *The Public Lands* (Madison: University of Wisconsin Press, 1968), p. 146.

28. Ibid., p. 149.

29. Fred A. Shannon, "Comment on 'The Railroad Land Grant Legend' in American History Texts," in Vernon Carstensen (ed.), *The Public Lands* (Madison: University of Wisconsin Press, 1968), p. 159.

30. Ibid., p. 158.

31. Greene, "Promised Land," p. 724.

32. Ibid.

33. Ibid., p. 725.

34. Shannon, "Comment on 'The Railroad Land Grant Legend,' " p. 159.

35. Hibbard, *History of the Public Land Policies*, p. 278.

36. Peter Barnes, "The Great American Land Grab," *New Republic*, June 5, 1971, pp. 19-23.

37. Hibbard, *History of the Public Land Policies*, p. 429.

Chapter 4   Federal Water Legislation and Practices

1. Phil Margolis, "Land Reclamation Laws: An Overview," *Rural America* 2, no. 7 (May 1977), p. 5.

2. Peter Barnes, "The Great American Land Grab," *New Republic*, June 5, 1971, p. 22; and E. Philip Leveen, "Natural Resource Development and State Policy: Origins and Significance of the Crisis in Reclamation," *Antipode* 11, no. 2 (1979), p. 62.

3. Barnes, "The Great American Land Grab," p. 23.

4. Paul S. Taylor, "The Law Says 160 Acres," *People and Land*, Summer 1973, p. 21.

5. Margolis, "Land Reclamation Laws," p. 5.

6. George L. Baker, "The Great Water Boondoggle," *The Progressive*, July 1978, p. 30.

7. Cited in ibid.

8. Cited in ibid.

9. Charles W. Howe and K. Williams Easter, *Interbasin Transfers of Water* (Baltimore, Md.: Johns Hopkins University Press, 1971), pp. 163–164.

10. Lynn T. Smith, "A Study of Social Stratification in Agricultural Sections of the U.S.," *Rural Sociology* 34, no. 4 (1969), pp. 496–509.

11. His letter appears in U.S., Congress, Senate, *Farmworkers in Rural America, 1971–1972, Hearings* before the Subcommittee on Migratory Labor of the Committee on Labor and Public Welfare, 92d Cong., 2d sess., 1972, part 3A, pp. 833–835 (hereinafter *Farmworkers in Rural America*).

12. Ibid.

13. Ibid., p. 833.

14. Baker, "The Great Water Boondoggle," p. 30; and Peter Barnes, "Water, Water for the Wealthy," *New Republic*, May 8, 1971, pp. 9–11.

15. George Ballis and Maia Sortor, "Water, Land and Power," *Congressional Record* 118, no. 80, 92d Cong., 2d sess. (May 17, 1972), p. 3.

16. "Bureau of Reclamation Project Areas," *National Land for People*, March 1979, p. 6.

17. Barnes, "Water, Water for the Wealthy," p. 10.

18. "New York Deserves $813 Billion Under Westlands Subsidy Formula," *National Land for People*, July 1978, back page.

19. Baker, "The Great Water Boondoggle," p. 30.

20. Ibid.

21. Joe Belden and Jackie Lundy, *Water and the West* (Washington, D.C.: Rural America, 1977).

22. "The 160-Acre Limit is Ditched Again," *People and Land*, Summer 1974, p. 6. For a grower's view, see Ralph M. Brody, general manager of the Westlands Water District in *Farmworkers in Rural America*, part 3C, pp. 1665–1673.

23. Cited in Baker, "The Great Water Boondoggle," p. 31.

24. Reported in *National Land for People*, February 1979.

25. Paul S. Taylor, "Mexican Irrigation and the 160-Acre Water Limitation," *California Law Review* 63, no. 3 (1975), p. 742.

26. Ibid., p. 743.

Chapter 5   This Land Is Not Our Land:
Who Owns Rural America?

1. David L. Ostendorf, "Food or Fuel?," *The Progressive*, April 1979, p. 45.

2. Peter Meyer, "Land Rush," *Harper's*, January 1979, p. 49.

3. Ibid.

4. Robert C. Fellmeth, *Politics of Land* (New York: Grossman Publishers, 1973), pp. 13–16.

5. Ibid., p. 10. How these corporations acquired their land holdings is described by George Ballis in U.S., Congress, Senate, *Farmworkers in Rural America, 1971–1972, Hearings* before the Subcommittee on Migratory Labor of the Committee of Labor and Public Welfare, 92d Cong., 2d sess., 1972, part 3A, pp. 727–730.

6. *The Austin-American Statesman* reported this information, which is cited in Meyer, "Land Rush," p. 49.

7. From a 1978 statement made by Senator Adlai E. Stevenson, Democrat from Illinois.

8. Ibid.

9. Meyer, "Land Rush," p. 49.

10. "Nebraska's New Land Barons," *People and Land*, Winter 1974, p. 22.

11. C. Barron McIntosh, "Forests Lieu Selections in the Sand Hills of Nebraska," *Annals of the Association of American Geographers* 64, no. 1 (March 1974), pp. 87-99.

12. Dan Looker, "Are Texans Buying Up the Land?" *New Land Review*, Fall 1976, p. 10.

13. Kemp Houck, *Concentrated Land Tenure in Kansas* (Lawrence: Kansas Farm Project, 1973), p. 10.

14. R. Barlowe and L. Libby, "Policy Choices Affecting Access to Farmland," in H. R. Guither (ed.), *Who Will Control U.S. Agriculture?*, North Central Regional Extension Publication 32 (Urbana-Champaign: University of Illinois, College of Agriculture, 1972), pp. 2-26.

15. C. V. Moore and C. W. Dean, "Industrialized Farming," in A. Gordon Ball and Earl O. Heady (eds.), *Size, Structure, and Future of Farms* (Ames: Iowa State University Press, 1972), pp. 214-231; and *1974 Census of Agriculture*.

16. B. Delworth Gardner and Rulon Pope, "How is Scale and Structure Determined in Agriculture?" *American Journal of Agricultural Economics* 60 (May 1978), p. 299.

17. Phil Margolis, "Foreign Land—Investors Underline Problems of Absentee Ownership," *Rural America* 3, no. 8 (July 1978), p. 6.

18. Charmaine Daniels, "U.S. and Foreign Investors Like Wall Street's New Blue Chip Category: Real Estate," *Maine Land Advocate* 6, no. 2 (March/April 1978), p. 4.

19. Letter to the editor, *The Progressive*, January 1979, p. 56.

20. Ed Schaefer, *Changing Character and Structure of American Agriculture: An Overview* (Washington, D.C.: General Accounting Office, 1978), p. 114. For a detailed account of foreign landlords in Nebraska, see Tom Graff, "Study of USDA Documents Reveals Nebraska's Foreign Landlords," *New Land Review*, Winter 1979/1980, pp. 1 and 3.

21. In Margolis, "Foreign Land," p. 6.

22. Data provided by the Emergency Land Fund, 836 Beecher Street, SW, Atlanta, Georgia.

23. Lester M. Salamon, "Family Assistance—The Stakes in the Rural South," *New Republic*, February 20, 1971, p. 17.

24. Manning Marable, "Black Agriculture in the Seventies," *In These Times*, January 25-31, 1978, p.6.

25. Salamon, "Family Assistance," pp. 17-18.

26. C. Scott Garber, "Where There's No Will, There's No Way," *Rural America* 3, no. 8 (July 1978), p. 3.

27. "Southerners Tackle Land Issue," *People and Land*, Summer 1974,

p. 16. For several specific examples, see Frank Sikora, "Black Land Loss Inquiries Flood Selma Lawyer," *Birmingham News*, August 6, 1978, p. 2B.

28. James S. Fisher, "Negro Farm Ownership in the South," *Annals of the Association of American Geographers* 63, no. 4 (December 1973), pp. 478–489.

29. NACLA, *Latin America and Empire Report* 2, no. 3 (March 1977), issue devoted to "The Apparel Industry Moves South."

30. Richard Peet, "Rural Inequality and Regional Planning," *Antipode* 7, no. 3 (1975), pp. 10–24.

31. Joe Persky, "Regional Colonialism and the Southern Economy," *Review of Radical Political Economy* 4 (Fall 1972), pp. 70–79.

32. Andre Gorz, "Colonialism, At Home and Abroad," *Liberation*, November 1971.

33. Meyer, "Land Rush," p. 49.

34. *Our Land and Water Resources*, USDA, Economic Research Service, Miscellaneous Publication No. 1290 (Washington, D.C.: Government Printing Office, 1974), p. 32.

35. "Biggest Farm Threat?" *The Country Today*, May 10, 1979, p. 1.

36. Center for Rural Affairs, *Newsletter*, August 1979.

Chapter 6   The Myth of Large-Scale Efficiency

1. *Changes in Farm Proudction and Efficiency*, USDA, Economic Research Service, Statistical Bulletin No. 548 (Washington, D.C.: Government Printing Office, 1963). Productivity is defined as output per unit input in one season.

2. Reduction of labor in agriculture has been rapid but actually has only shifted. John and Carol Steinhard argued that "there has been no reduction in [agriculture and food labor] at all—only a change in what workers do. Yesterday's farmer is today's canner, tractor mechanic and fast food car hop." Cited in W. Clark, "U.S. Agriculture is Growing Trouble as well as Crops," *Smithsonian* 5, no. 10 (January 1975), p. 62.

3. Luther G. Tweeten, "Theories Explaining the Persistence of Low Resource Returns in a Growing Farm Economy," *American Journal of Agricultural Economics* 51, no. 4 (1969), p. 809.

4. Ibid.

5. Ibid.

6. Michael Perelman, *Farming for Proift in a Hungry World* (Montclair, N.J.: Allanheld, Osmun & Co., 1977), p. 96, table 10-1.

7. Frances Moore Lappe and Joseph Collins, *Food First: Beyond the Myth of Scarcity* (Boston: Houghton Mifflin Co., 1977), p. 145.

8. Jim Hightower, *Eat Your Heart Out: Food Profiteering in America* (New York: Crown Publishers, 1975), p. 60.

9. Michael Perelman, "Efficiency and Agriculture," in U.S., Congress, Senate, *Farmworkers in Rural America, 1971-1972, Hearings* before the Subcommittee on Migratory Labor of the Committee on Labor and Public Welfare, 92d Cong., 2d sess., 1972, part 3A, pp. 1084–1086.

10. The concept of economies of size is more appropriate than that of economies of scale. Returns to scale refer to the impact on farm output of a given change in the level of *all* resources in fixed proportion. In reality, expansion of farming units does not entail an equal percentage increase in all resources. Instead, farmers increase the proportion of capital to labor and of variable capital to fixed capital.

11. Most cost-size studies are not based on statistical analysis of actual performance of farms of various size but rather on hypothetical or synthetic models, which are intended to be representative of various farm sizes. For a critique of these synthetic studies, see the final report of the Small Farm Viability Project, *The Family Farm in California* (November 1977), Technology Task Force report, appendix A.

12. J. Patrick Madden, *Economies of Size in Farming*, USDA, Economic Research Service, Agricultural Economic Report No. 107 (Washington, D.C.: Government Printing Office, 1967), p. 54. Madden reached this same conclusion five years later in J. Patrick Madden and Earl J. Partenheimer, "Evidence of Economies and Diseconomies of Farm Size," in A. Gordon Ball and Earl O. Heady (eds.), *Size, Structure, and Future of Farms* (Ames: Iowa State University Press, 1972), pp. 91–107.

13. B. C. French, "The Analysis of Productive Efficiency in Agricultural Marketing: Models, Methods, and Progress," in Lee R. Martin (ed.), *A Survey of Agricultural Economics Literature* (Minneapolis: University of Minnesota Press, 1977), pp. 131–132.

14. Philip M. Raup, "Economies and Diseconomies of Large-Scale Agriculture," *American Journal of Agricultural Economics* 51 (1969), p. 1279.

15. Madden and Partenheimer, "Evidence of Economies and Diseconomies of Farm Size," pp. 96–97.

16. Ibid., p. 97.

17. Madden, *Economies of Size in Farming*, p. 70.

18. D. K. Britton and Berkeley Hill, *Size and Efficiency in Farming* (Lexington, Mass.: Lexington Books, 1975), pp. 9–10.

19. J. Bruce Hottel and Robert D. Reinsel, *Returns to Equity Capital by Economic Class of Farm*, USDA, Economic Research Service Report No. 347 (Washington, D.C.: Government Printing Office, 1976), p. 14.

20. B. F. Stanton, "Perspective on Farm Size," *American Journal of Agricultural Economics* 60, no. 5 (1978), p. 735.

21. Don Villarejo, "The Political Economy of Agribusiness in California," *Congressional Record* 122, no. 150, 94th Cong., 2d sess. (September 30, 1976), p. E5343.

22. Ibid.

23. French, "Analysis of Productive Efficiency," p. 150, cites evidence that smaller capital-intensive agricultural marketing and processing plants show significant gains in economies of size at 80 percent of the largest plants.

24. Virgil L. Christian, Jr., and Carl C. Erwin, "Agriculture," in Ray Marshall and Virgil L. Christian, Jr. (eds.), *Employment of Blacks in the South* (Austin: University of Texas, 1978), p. 58.

25. In the absence of empirical studies on increased efficiency by means

of intensifying production, much of this discussion must remain theoretical. For a further discussion of these ideas, see William Bunge, *Urban Nationalism* (Ottawa: Ministry of State for Urban Affairs, 1973), pp. 61–67, particularly figure 7, p. 66.

26. *The Changing Structure of U.S. Agribusiness and Its Contribution to the National Economy* (Washington, D.C.: U.S. Chamber of Commerce, 1974), p. 6.

27. Quoted by Perelman, *Farming for Profit*, p. 118.

28. Cited in Britton and Hill, *Size and Efficiency in Farming*, p. 118.

29. W. Burt Sundquist, "Scale Economies and Management Requirements," in A. Gordon Ball and Earl O. Heady (eds.), *Size, Structure, and Future of Farms* (Ames: Iowa State University Press, 1972), p. 85.

30. Peter Emerson, *Public Policy and the Changing Structure of American Agriculture*, Congressional Budget Office (Washington, D.C.: Government Printing Office, 1978), p. 29 and table 4 on p. 30.

31. See French, "Analysis of Productive Efficiency," p. 94.

32. Kenneth Boulding, "In Praise of Inefficiency," *Graduate Women* 73, no. 4 (1979), p. 29.

33. Wendell Berry, *The Unsettling of America: Culture and Agriculture* (New York: Avon Books, 1977), p. 7.

Chapter 7   The Business of Agribusiness

1. J. H. Davis and R. A. Goldberg, *A Concept of Agribusiness* (Boston: Alpine Press for Harvard University, 1957), p. 9.

2. *The Changing Structure of U.S. Agribusiness and Its Contribution to the National Economy* (Washington, D.C.: U.S. Chamber of Commerce, 1974), p. 9.

3. *Agribusiness Stock Ownership Directory* (New York: Corporate Data Exchange, 1978) examines the 222 leading agribusiness companies for the major shareholders voting the companies' stock, the investment banks underwriting the companies' securities, and the law firms and lobbyists providing key legal services. *The AgBiz Tiller* (San Francisco Study Center, P.O. Box 5646, San Francisco, CA 94101) preview issue lists the revenues, assets, profits, and profitability of large U.S. corporations involved in agribusiness. For a description of how food conglomerates gobble up each other and consumers, see Daniel Zwerdling, "The Food Monsters," *The Progressive*, March 1980, pp. 16–27.

4. *Changing Structure of U.S. Agribusiness*, p. 13. The USDA excludes expenditures for new buildings and machinery from annual production costs, considering them fixed costs.

5. Cited by Jim Hightower, *Eat Your Heart Out: Food Profiteering in America* (New York: Crown Publishers, 1975), p. 140.

6. Nearly three-quarters of the corn planted in the U.S. is of only 6 varieties out of the 197 varieties adapted to growing in this country. Dan McCurry, "Tarnished Gold," *New Land Review*, Winter 1978, pp. 12 and 11.

7. Tyler Sutton, "Ammonia Industry Profits Engulf Farm Income," *New Land Review*, Winter 1977, p. 5.

8. "Riding the Farm Boom," *Business Week*, October 27, 1973, p. 78.

9. Hightower, *Eat Your Heart Out*, p. 141.

10. "Profits: Better Than Expected," *Business Week*, May 11, 1974, p. 69.

11. Paul D. Scanlon, "FTC and Phase II: The McGovern Papers," *Antitrust Law and Economic Review* 5, no. 3 (1972), table 1, pp. 33-36.

12. *Changing Structure of U.S. Agribusiness*, p. 2.

13. Mary Hamblin, "Bank Lending to Agriculture: An Overview," *Monthly Review*, Kansas Federal Reserve District, November 1975, p. 14.

14. *Where Have All The Bankers Gone?* (Walthill, Neb.: Center for Rural Affairs, 1977), p. 9.

15. *1978 Handbook of Agricultural Charts* (Washington, D.C.: Government Printing Office, 1978), p. 20.

16. *1974 Census of Agriculture*, Statistics by Subject, vol. 2, part 2, table 25, p. III-14.

17. Emanuel Melichar, "Some Current Aspects of Agricultural Finance and Banking in the United States," *American Journal of Agricultural Economics* 59 (December 1977), p. 970.

18. Divizich's testimony appears in U.S., Congress, Senate, *Farmworkers in Rural America, 1971-1972, Hearings* before the Subcommittee on Migratory Labor of the Committee on Labor and Public Welfare, 92d Cong., 2d sess., 1972, part 3B, pp. 1222-1223.

19. Cited in "NFO Sees Drop in Dairy Support Price by Fall," *The Country Today*, March 28, 1979, p. 5.

20. "Perpetual Farm Debt Justified," *The Country Today*, February 7, 1979, pp. 1 and 10.

21. Banks are destroying family farmers directly by investing in farmland and turning farmers into tenants. See Cathy Lerza, "Corporate Sharecropping," *Environmental Action* 8, no. 20 (1977), pp. 71-72; George Ballis, "Investors Eye California Farmland," *Rural America* 2, no. 9 (July–August 1977), p. 4; John Grossman, "The Bank versus the Family Farmers," *Chicago Magazine*, November 1977, pp. 228-241; and U.S., Congress, House, *Ag-Land Trust Proposal, Hearings* before the Subcommittee on Family Farms, Rural Development, and Special Studies of the Committee on Agriculture, 95th Cong., 1st sess., 1977, serial No. 95-A.

22. "The Menace of Agribusiness," *Universal Times*, January 1976, p. 1.

23. Leonard R. Kyle, W. B. Sundquist, and Harold D. Guither, "Who Controls Agriculture Now?—The Trends Underway," in *Who Will Control U.S. Agriculture?* North Central Regional Extension Publication 37 (Urbana-Champaign: University of Illinois, College of Agriculture, August 1972), p. 9.

24. Cliff Conner, "Hunger: U.S. Agribusiness and World Famine," *International Socialist Review* 35, no. 8 (1974), pp. 20-31.

25. *1978 Handbook of Agricultural Charts*, p. 4.

26. Ibid., p. 7.

27. *Welcome to Monfort's Feed Lots*, company brochure, 1970.

28. For examples of vertically integrated agricultural firms in the South, see Mark Pinsky, "The New Plantations of the South," *Rural America* 2, no. 11 (1977), pp. 6-7.

29. Larry Casalino, "This Land is Their Land," *Ramparts*, July 1972, pp. 31–36.

30. Ray A. Goldberg, "U.S. Agribusiness Breaks Out of Isolation," *Harvard Business Review*, May–June 1975, pp. 81–95.

31. R. L. Mighell and W. S. Hoffnagle, *Contract Production and Vertical Integration in Farming, 1960 and 1970*, USDA, Economic Research Service Report No. 479 (Washington, D.C.: Government Printing Office, 1970).

32. "The Menace of Agribusiness," pp. 1, 4, and 11.

33. A. V. Krebs, *Major U.S. Corporations Involved in Agribusiness* (Washington, D.C.: Agribusiness Accountability Project, 1973); and Philip Raup, *Nature and Extent of the Expansion of Corporations in American Agriculture,* Staff Report P75-8 (St. Paul: University of Minnesota, Department of Agricultural and Applied Economics, April 1975).

34. The founding family of Cargill owns 85 percent of the company's common stocks and nonfamily executives share the rest. See Hubert Kay, "The Two-Billion Dollar Company That Lives by Cents," *Fortune*, December 1965, pp. 167–202.

35. Ed Schaefer, *Changing Character and Structure of American Agriculture: An Overview* (Washington, D.C.: General Accounting Office, 1978), pp. 109–111.

36. C. V. Moore and J. H. Snyder, "Corporate Farming in California," *California Agriculture*, March 1970, p. 70.

37. "Farm Boom: Water, Water, Everywhere," *Newsweek*, May 20, 1974, pp. 83–84; also, *AgBiz Tiller*, no. 1 (August 1976), pp. 1–7.

38. Richard D. Rodefeld, "A Reassessment of the Status and Trends in 'Family' and 'Corporate' Farms in U.S. Society," *Congressional Record*, 93d Cong., 1st sess. (May 31, 1973), p. S17622.

39. Ibid.

40. *Corporations Having Agricultural Operations*, USDA, Economic Research Service Report No. 142 (Washington, D.C.: Government Printing Office, 1968).

41. Philip M. Raup, "Corporate Farming in the U.S.," *Journal of Economic History* 33, no. 1 (1973), p. 275.

42. Ronald O. Aines, "Linkages in Control and Management with Agribusiness," in A. Gordon Ball and Earl O. Heady (eds.), *Size, Structure, and Future of Farms* (Ames: Iowa State University Press, 1972), p. 195.

43. Cited in a press release from Governor Wendell Anderson of Minnesota in 1971.

44. *1974 Census of Agriculture*. For a review of corporate farming in the West, see James J. Parsons, "Corporate Farming in California," *Geographical Review* 67, no. 3 (July 1977), pp. 354–357.

45. Philip Raup, *What Policies Should We Have Toward Corporations in Farming?* (St. Paul: University of Minnesota, Department of Agricultural and Applied Economics, December 1969), p. 4.

46. Charles V. Moore and Gerald W. Dean, "Industrialized Farming," in A. Gordon Ball and Earl O. Heady (eds.), *Size, Structure, and Future of Farms* (Ames: Iowa State University Press, 1972), pp. 216–218.

47. Ibid., p. 217.

48. Schaefer, *Changing Character and Structure of American Agriculture*, p. 76.

49. For samples of grower contracts, see *Farmworkers in Rural America*, pp. 1241-1260; and Hightower, *Eat Your Heart Out*, pp. 255-259.

50. Testimony presented before Senator Gaylord Nelson's U.S. Senate Subcommittee on Small Business hearings on *Corporate Giantism and Food Prices* (Washington, D.C.: Government Printing Office, 1973).

51. Dale O. Anderson, "Economic Means of Farm Groups," in A. Gordon Ball and Earl O. Heady (eds.), *Size, Structure, and Future of Farms* (Ames: Iowa State University Press, 1972), pp. 353-372.

52. *Feedstuff*, December 12, 1970, p. 4.

53. For a detailed analysis of the integrated chicken industry, see Stephen Singular, "Brave New Chickens," *New Times*, April 19, 1977, pp. 42-55.

54. Don Paarlberg, *American Farm Policy* (New York: John Wiley & Sons, 1964), p. 40.

55. Sheldon L. Greene, "Corporate Feudalism in Rural America," in *Farmworkers in Rural America*, part 3C, p. 2034.

56. *Science for the People*, December 1978, p. 4.

57. R. Nikolitch, "Family-Operated Farms: Their Compatability with Technological Advance," *American Journal of Agricultural Economics* 15 (1969), p. 541.

58. Hightower, *Eat Your Heart Out*, p. 164.

59. For an excellent review of the negative consequences of contract farming for family farmers, see Frances Moore Lappe and Joseph Collins, *Food First: Beyond the Myth of Scarcity* (Boston: Houghton Mifflin Co., 1977), pp. 272-278.

60. "Strikers Take Aim at Canners in Ohio," *Rural America* 4, no. 3 (1979), p. 10. For a conventional analysis of contract farming, see Bruce W. Marion (ed.), *Coordination and Exchange in Agricultural Subsectors*, North Central Research Publication No. 228 (Madison: University of Wisconsin, College of Agricultural and Life Sciences, 1976).

Chapter 8   Tax-Loss Farming

1. Jeanne Dangerfield, "Sowing the Till," *Congressional Record* 119, no. 74, 93d Cong., 1st sess. (May 16, 1973), pp. S9247-S9255.

2. "Senate Debates Chickens, Strawberries, and Annuities," *People and Taxes*, June 1977, pp. 4, 5, and 8.

3. Harold F. Breimyer et al., "How Federal Income Tax Rules Affect Ownership and Control of Farming," in *Who Will Control U.S. Agriculture?*, North Central Regional Extension Publication 37 (Urbana-Champaign: University of Illinois, College of Agriculture, July 1974), p. 3.

4. "Farm Size and the 1978 Tax Law: Help to Get Big or Pressure to Get Out," *Newsletter*, Center for Rural Affairs, March 1979, last two pages.

5. Peter Barnes, "The Farmer-Doctor," *New Republic*, September 2, 1972, p. 14.

6. "Farm Size and the 1978 Tax Law."

7. Breimyer et al., "How Federal Income Tax Rules Affect Ownership and Control," p. 2.

8. Dangerfield, "Sowing the Till."

9. Charles Davenport, "Tax Loss Farming by Syndicates and Corporations" (Paper delivered at the First National Conference on Land Reform, San Francisco, April 27, 1973), p. 4.

10. Thomas A. Carlin and W. Fred Woods, *Tax Loss Farming*, USDA, Economic Research Service Report No. 546 (Washington, D.C.: Government Printing Office, 1974).

11. "Tax Loss is Our Loss," *People and Land*, Summer 1974, p. 7. Angus McDonald analyzed the 1970 Internal Revenue Service data.

12. Chuck Hassebrook, "The Lure of Land Speculation," *New Land Review*, Summer 1979, p. 12.

13. Breimyer et al., "How Federal Income Tax Rules Affect Ownership and Control," p. 4.

14. "Tax Shelters," *People and Taxes* 6, no. 3 (March 1978), pp. 9–10.

Chapter 9 Federal Farm Programs

1. Cited in George Melloan, "Time to Phase Out Farm Subsidies?" *Wall Street Journal*, May 4, 1972, p. 4.

2. Leroy Quance and Luther G. Tweeten, "Policies, 1930–1970," in A. Gordon Ball and Earl O. Heady (eds.), *Size, Structure, and Future of Farms* (Ames: Iowa State University Press, 1972), p. 36.

3. Ed Schaefer, *Changing Character and Structure of American Agriculture: An Overview* (Washington, D.C.: General Accounting Office, 1978), p. 59.

4. John R. Tarrant, *Agricultural Geography* (New York: John Wiley & Sons, 1974), p. 205. Also see Don F. Hadwiger and Ross B. Talbot, *Pressures and Protests* (San Francisco: Chandler Publishing Co., 1965), p. 21; and Harland I. Padfield, "Agrarian Capitalists and Urban Proletariat," in W. G. McGinnies, B. J. Goldman, and P. Paylore (eds.), *Food, Fiber, and the Arid Lands* (Tucson: University of Arizona Press, 1971), p. 44.

5. Joshua Bernhardt, *The Sugar Industry and the Federal Government* (Washington, D.C.: Sugar Statistics Service, 1948), p. 202.

6. Neal R. Peirce, *The Megastates of America* (New York: W. W. Norton & Co., 1972), p. 79.

7. Karl Raitz, "Federal Agriculture Policy and Rural America," *Antipode* 7, no. 3 (December 1975), p. 5.

8. Luther Tweeten, "Commodity Programs for Agriculture," in Vernon Ruttan et al. (eds.), *Agricultural Policy in an Affluent Society* (New York: W. W. Norton & Co., 1969), p. 111.

9. Raitz, "Federal Agricultural Policy," pp. 5–6.

10. Howard F. Gregor, *Geography of Agriculture* (Englewoods Cliffs, N. J.: Prentice-Hall, 1970), p. 86.

11. James T. Bonnen, "The Distribution of Benefits From Selected U.S.

Farm Programs," in *Rural Poverty in the United States*, report by the President's National Advisory Commission on Rural Poverty (Washington, D.C.: Government Printing Office, 1968), p. 486, table 18, and p. 483.

12. Cited by Robert L. Gnaizda, "Is it 'More Blessed to Give than to Receive'?—A Study of Federally Subsidized Grower Attitudes Toward Federal Assistance to the Poor," in U.S., Congress, Senate, *Farmworkers in Rural America, 1971-1972, Hearings* before the Subcommittee on Migratory Labor of the Committee on Labor and Public Welfare, 92d Cong., 2d sess., 1972, part 5B, p. 4039 (hereinafter *Farmworkers in Rural America*).

13. Tarrant, *Agricultural Geography*, p. 207.

14. Leslie Hewes, *The Suitcase Farming Frontier* (Lincoln: University of Nebraska Press, 1973), p. 144.

15. Neal R. Peirce, *The Deep South States in America* (New York: W. W. Norton & Co., 1974), p. 190.

16. Ibid., p. 198.

17. Raitz, "Federal Agricultural Policy," p. 7.

18. *Farmworkers in Rural America*, part 3C, pp. 1662-1663.

19. Bonnen, "Distribution of Benefits." The distribution of benefits is far more critical than the total dollar value of the programs, which varies from year to year, crop to crop, and program to program.

20. *New Land Review*, Winter 1978, p. 4.

21. Charles L. Schultze, *The Distribution of Farm Subsidies: Who Gets the Benefits* (Washington, D.C.: Brookings Institution, 1971), p. 17; italics added.

22. Ibid.

23. Ibid., p. 18.

24. Ibid., p. 29.

25. Stanley D. Brunn, *Geography and Politics in America* (New York: Harper and Row, 1974), p. 334.

26. Paul Findley, "Farm Payments over $25,000," *Congressional Record*, 91st Cong., 2d sess. (March 26, 1970), p. 9632.

27. Brunn, *Geography and Politics in America*, p. 334.

28. "Farm Subsidy Payments Run Into Millions Of Dollars," *Congressional Quarterly*, June 12, 1970, p. 1549.

29. Brunn, *Geography and Politics in America*, p. 336.

30. "Farm Subsidy Payments Run Into Millions of Dollars."

31. *Farmworkers in Rural America*, part 3A, p. 719.

32. Schultze, *Distribution of Farm Subsidies*, pp. 1-2.

33. Howard Wachtel, "Looking at Poverty from a Radical Perspective," *Review of Radical Political Economics* 3, no. 3 (Summer 1971), p. 12.

34. For an appreciation of the magnitude and intensity of hunger in the U.S., see "From 'Hunger U.S.A. Revisited,' " *New South*, Fall 1972, pp. 35-49.

35. *New York Times*, magazine section, March 7, 1971.

36. Schultze, *Distribution of Farm Subsidies*, p. 3.

37. Ibid., p. 44.

38. Jim Chapin, "The Rich are Different. . . ," *Newsletter of the Democratic Left*, November 1976, p. 3.

39. A Paulsen, "Payment Limitations: The Economic and Political Feasibility," *American Journal of Agricultural Economics* 51 (1969), p. 124.

40. Ibid.

41. Brunn, *Geography and Politics in America*, p. 336. Districts most likely to benefit from federal irrigation projects are also overrepresented in Congress. "Since the 1930s, roughly 90% of the appropriate Senate committees and 70% of the comparable House committees have been from the 17 Western states." E. Philip Leveen, "Natural Resource Development and State Policy: Origins and Significance of the Crisis in Reclamation," *Antipode* 11, no. 2 (1979), p. 70.

42. Brunn, *Geography and Politics in America*, p. 337.

43. "Congress Nears Completion of Farm and Food Bill," *Rural America* 2, no. 10 (September 1977), p. 1.

44. Ibid.

45. See R. W. Bartlett, "What Government Price Policies for the Dairy Industry?" *Illinois Agricultural Economics,* July 1976, pp. 17–22, for an overview of the shortcomings in the dairy programs for public policy and consumer prices.

46. David R. Fronk, *Farm Size and Regional Distribution of the Benefit under Federal Milk Market Regulation* (Washington, D.C.: Federal Trade Commission, Bureau of Economics, 1978), p. 1.

47. Ibid., p. 10.

48. Ibid., p. 20.

49. Ibid., p. 21.

50. Frank Wright, "Dairy Industry's Political War Chest Brings Payoff," *Minneapolis Tribune,* March 28, 1971, pp. 1A and 10A. The following members of Congress helped the lobbying campaign by cosponsoring legislation; if passed, this law would have forced Hardin to raise the milk support price. Others, such as Belcher, did not put their name on the legislation but pressured the administration personally. Among the congressional allies of the milk lobby were some of the most powerful and well-known persons on Capitol Hill: House Speaker Carl Albert, D-Okla.; Rep. W. R. Poage, D-Texas, Chairman of the House Agriculture Committee; Rep. Page Belcher, R-Okla., senior minority member of the Agriculture Committee; Sen. Hubert Humphrey, D-Minn., a member of the Senate Agriculture Committee; Sen. Edmund Muskie, D-Maine, the leading contender for the 1972 Democratic presidential nomination; Sen. Harold Hughes, D-Iowa, a darkhorse prospect for that nomination; Sen. William Proxmire, D-Wisc., a ranking member of the Senate Agriculture Appropriations Subcommittee; and Sen. Gale McGee, D-Wyo., chairman of that subcommittee.

51. See Brooks Jackson, "Milking the White House," *New Republic,* January 26, 1974, p. 17, for events and amounts spent on lobbying.

52. Brooks Jackson, "Bribery and Contributions," *New Republic,* December 21, 1974, p. 14.

53. Brooks Jackson, "Milk Money," *New Republic,* August 10 and 17, 1974, p. 14.

54. Quoted in Nicholas Wade, "Thirty-one Favors," *New Republic,* November 5, 1977, p. 16.

55. Emily Yoffe, "Two Milk Duds," *New Republic*, July 23, 1977, pp. 9–10.

56. Cited in Kennedy P. Maize, "Calling Dr. Kahn," *New Republic*, December 9, 1978, p. 15.

57. Senator Gaylord Nelson's *Newsletter*, March 1977.

58. *Who Will Farm?* (New York: United Presbyterian Church in the U.S.A., Church and Society Program Area, 1978), p. 15.

59. Virgil L. Christian, Jr., and Carl C. Erwin, "Agriculture," in Ray Marshall and Virgil L. Christian, Jr. (eds.), *Employment of Blacks in the South* (Austin: University of Texas Press, 1978), pp. 68–69.

60. William C. Payne, "Implementing Federal Nondiscrimination Policies in the Department of Agriculture, 1964–1976" (Paper presented at the Agricultural Policy Symposium of the Policy Studies Organization, Washington, D.C., July 26, 1977), pp. 12–13.

61. Christian and Erwin, "Agriculture," p. 72.

62. *Who Will Farm?*

Chapter 10   The Land-Grant College System

1. Lauren Soth, "Tides of National Development," in U.S., Congress, Senate, *Farmworkers in Rural America, 1971-1972, Hearings* before the Subcommittee on Migratory Labor of the Committee on Labor and Public Welfare, 92d Cong., 2d sess., 1972, part 4B, p. 2902 (hereinafter *Farmworkers in Rural America*).

2. Ibid., p. 2903.

3. H. R. Fortmann, "Colleges of Agriculture Revisited," *Farmworkers in Rural America*, p. 3033.

4. Jim Hightower, *Hard Tomatoes, Hard Times* (Cambridge, Mass.: Schenkman Publishing Co., 1973), p. 9.

5. Leonard D. White, "The Department of Agriculture," in Louis H. Douglas (ed.), *Agrarianism in American History* (Lexington, Mass.: D. C. Heath & Co., 1969), pp. 65–66.

6. Ibid., p. 68.

7. *Administrative Manual for the Hatch Act as Amended*, USDA, Cooperative State Research Service, Agricultural Handbook No. 381 (Washington, D.C.: Government Printing Office, July 1970); italics added.

8. Hightower, *Hard Tomatoes, Hard Times*, p. 14.

9. *A People and A Spirit*, report of the Joint USDA-NASULGC Study Committee (Ft. Collins: Colorado State University, November 1968), p. 18; italics added.

10. *Farmworkers in Rural America*, part 4A, p. 2163.

11. Ibid., part 4B, p. 2770.

12. Ibid., part 4A, p. 2187.

13. Fred Schmidt and David Clavelle, *Issues on Agricultural Research* (Washington, D.C.: Rural America, 1977).

14. *Rural America* 4, no. 5 (May 1979), p. 8.

15. Schmidt and Clavelle, *Issues on Agricultural Research*, based on a U.S. General Accounting Office study.

16. Hightower, *Hard Tomatoes, Hard Times*, p. 27.

17. H. J. Hodgson, "Scope and Character of Public Plant Breeding in the U.S.," *Agricultural Science Review* 9, no 4 (1971), pp. 36–43.

18. "Wilson's Seed Money," *Time*, May 29, 1972, p. 78.

19. Andrew Schmitz and David Seckler, "Mechanized Agriculture and Social Welfare: The Case of the Tomato Harvester," *American Journal of Agricultural Economics* 52, no. 4 (1970), p. 570.

20. Ibid., p. 575. Nineteen farm workers, represented by California Rural Legal Assistance, filed suit in February 1979 against the Board of Regents of the University of California. The suit charged that taxpayers' dollars were being used to conduct agricultural research that forced farm workers out of work. The suit further contended that research in highly mechanized, capital-intensive farming methods was, in effect, an illegal subsidy of large agribusiness concerns. See "Cloning Asparagus at the Public's Expense," *Rural America* 4, no. 5 (May 1979), p. 8. Another challenge to mechanization research is being made by the United Farm Workers Union, which is working with California Assemblyman Art Torrez to introduce a bill requiring a "social-impact statement" as a precondition for using machines that would displace farm workers. The UFW estimates that 120,000 jobs will be lost to mechanization over the next ten years.

21. William H. Friedland and Amy Barton, *Destalking the Wily Tomato: A Case Study in Social Consequences in California Agricultural Research*, Research Monograph No. 15 (Santa Cruz: University of California, 1975).

22. Paul Barnett, "Tough Tomatoes," *The Progressive*, December 1977, p. 33.

23. "Growers Prefer Machines to UFW Members," *Dollars and Sense*, March 1978, pp. 14–16.

24. Barnett, "Tough Tomatoes," p. 32.

25. Ibid., p. 34.

26. Ibid.

27. *Farmworkers in Rural America*, part 4A, p. 2166.

28. The impact of mechanization has been different for tobacco and cotton. Since cotton allotments under the federal allotment program were transferable—while tobacco allotments could not be separated from the land to which they were attached—large acreages could be assembled more easily in cotton than in tobacco production. Hence, black and white tobacco growers have not experienced as great a disadvantage in regard to farm size as cotton growers. In fact, between 1969 and 1974 tobacco farms, which received 50 percent or more of their total agricultural value from tobacco, increased by 6 percent, while cotton farms decreased 24 percent.

29. *Farmworkers in Rural America*, part 4A, p. 2173.

30. Ibid., part 1, p. 166.

31. Robert C. McElroy, "Manpower Implications of Trends in the Tobacco Industry," in ibid., part 4A, p. 2184.

32. Ibid.

33. Ibid., p. 2186.

34. *Farmworkers in Rural America*, part 4B, pp. 2773–2774.

35. Louis H. Douglas (ed.), *Agrarianism in American History* (Lexington, Mass.: D. C. Heath & Co., 1969), pp. 76–77.

36. Ibid.

37. William Payne, "The Negro Land-Grant Colleges," *Civil Rights Digest* 3, no. 2 (Spring 1970), p. 15.

38. Ibid.

39. Peter Schuck, "Black Land-Grant Colleges: Separate and Still Unequal," in *Farmworkers in Rural America*, part 4A, p. 2342.

40. *Farmworkers in Rural America*, part 4A, p. 2332.

41. Ibid., p. 2333.

42. Ibid. State aid to black schools is particularly critical because these schools have traditionally received a greater portion of their income from state aid than from federal sources. A report by the National Association of State Universities and Land-Grant Colleges showed that sixteen white land-grant colleges received $450 million in state appropriations—*almost nine times* the $52.3 million received by black schools in the same states. The federal and state imbalances cannot be explained by relative school enrollments, as the white land-grant colleges had only 5.5 times more students than the black land-grant schools in the same states. See Payne, "The Negro Land-Grant Colleges," p. 15.

43. Peter Schuck, "Black Land-Grant Colleges," p. 2353.

44. *Farmworkers in Rural America*, part 4B, p. 2769.

45. Ibid., part 4A, p. 2337.

46. Virgil L. Christian, Jr., and Carl C. Erwin, "Agriculture," in Ray Marshall and Virgil L. Christian, Jr. (eds.), *Employment of Blacks in the South* (Austin: University of Texas, 1978), p. 41.

47. Ibid., p. 44.

48. *Farmworkers in Rural America*, part 4A, pp. 2315 and 2341.

49. Ibid., part 4B, p. 2765.

50. Ibid., part 4A, p. 2333.

51. Ibid., p. 2337.

52. Hightower, *Hard Tomatoes, Hard Times*, p. 31. Also see Harold Lasswell, "From Fragmentation to Configuration," *Policy Sciences* 2, no. 4 (December 1971), p. 442.

53. Hightower, *Hard Tomatoes, Hard Times*, p. 28.

Chapter 11   The Plight of Seasonal Farm Families

1. For marital property rights of women, see Betty Gordon Becton and Betty Belk Moorhead, "What ERA Will Do for Family Law," *Graduate Women* 73, no. 4 (July/August 1979), pp. 32–35.

2. Robert Coles and Tom Davey, "Working With Migrants," *New Republic*, October 16, 1976, p. 15.

3. Ibid., p. 14.

4. For Paul S. Taylor's excellent review of the origins and growth of migratory seasonal labor in agriculture, see U.S., Congress, Senate, *Farmworkers in Rural America, 1971–1972, Hearings* before the Subcommittee on Migratory Labor of the Committee on Labor and Public Welfare, 92d Cong., 2d sess., 1972, part 5B, pp. 3892–3916 (hereinafter *Farmworkers in Rural America*).

5. Joan London and Henry Anderson, *So Shall Ye Reap* (New York: Thomas Y. Crowell, 1970), pp. 7–8.

6. Ibid., p. 14. See also Richard B. Craig, *The Bracero Program* (Austin: University of Texas Press, 1971).

7. Ernesto Galarza, in *Farm Workers and Agri-business in California, 1947–1960* (Notre Dame, Indiana: University of Notre Dame Press, 1977), pp. 215–216, documents this for California.

8. Ibid., pp. 226–227.

9. Ibid., p. 238.

10. London and Anderson, *So Shall Ye Reap*, p. 10.

11. For a full discussion of the enforcement and attacks on the bracero program, see Galarza, *Farm Workers and Agri-business in California*, chapter 4, pp. 203–276.

12. "Undocumented Workers," *In These Times*, 1978.

13. Ibid.

14. Sheldon L. Greene, "Illegal Alien Labor In California Agriculture: An Indirect Federal Subsidy," *Farmworkers in Rural America*, part 3C, pp. 1991–2027.

15. Alejandro Portes, "Return of the Wetback," *Society* 2, no. 3 (March/April 1974), p. 45.

16. Gino Germani, *Marginality* (New Jersey: Transaction Books, 1979).

17. London and Anderson, *So Shall Ye Reap*, pp. 15–16.

18. Quoted in Rius, *Marx for Beginners* (New York: Pantheon Books, 1976), p. 97.

19. Ernesto Galarza, *Merchants of Labor: The Mexican Bracero Story* (Santa Barbara, Calif.: McNally & Loftin, 1978), part 4, pp. 121–182.

20. Carey McWilliams, *Factories in the Fields* (Santa Barbara, Calif.: Peregrine Publishers, 1971), chaps. 14 and 15. The saga of how sixty apple growers in Virginia and growers in eleven other states used the courts to import almost 5,000 Jamaican pickers, while more than 100,000 unemployed domestic agricultural workers and thousands of unemployed urban teenagers from Washington, D.C., were bypassed against the wishes of the Department of Labor, is told by John Egerton in "As Jamaican as Apple Pie," *The Progressive*, December 1977, pp. 37–40; and Dorothy McGhee, "Apple Picker Blues," *New Republic*, October 29, 1977, pp. 15–16.

21. L. H. Fisher, *The Harvest Labor Market in California* (Cambridge, Mass.: Harvard University Press, 1953), p. 110.

22. McWilliams, *Factories in the Fields*, pp. 179–180.

23. *Where Have All the Farm Workers Gone?* Research Report No. 1 (Washington, D.C.: Rural America, 1977), p. x.

24. Ibid., p. ix.

25. Ibid., pp. 3 and 60.

26. See Philip Sorensen's statement in *Farmworkers in Rural America*, part 1, p. 108.

27. Stephen H. Sosnick, *Hired Hands: Seasonal Farm Workers in the United States* (Santa Barbara, Calif.: McNally & Loftin, 1978), p. 10.

28. James C. Harrington, "Mexican-Americans Seek a Just Place," *Civil Liberties*, no. 318 (July 1977), pp. 1 and 4.

29. Jane Kay, "Child Workers Stir Age-Old Labor Conflict," *Arizona Daily Star*, March 7, 1976, sec. B, p. 1.

30. "DOL Ignores Pesticide Warnings, Puts Children to Work in the Fields," *Rural America* 4, no. 1 (January 1979), pp. 1 and 10.

31. Sosnick, *Hired Hands*, p. 39.

32. "Many Farm Workers Paid Below Legal Minimum," *San Jose Mercury*, October 21, 1971, reprinted in *Farmworkers in Rural America*, part 3C, pp. 1794–1795.

33. Sosnick, *Hired Hands*, p. 40.

34. Robert Gnaizda, "The Federally-Funded, Grower-Oriented Farm Labor Service: A Quarter of a Billion Dollar Gift For Lawbreakers," in *Farmworkers in Rural America*, part 3C, p. 1977. For substantiation of the quote, refer to the whole article, pp. 1975–1984.

35. Sosnick, *Hired Hands*, p. 23.

36. *Farmworkers in Rural America*, part 3A, p. 1000.

37. *Farm Labor*, USDA (Washington, D.C.: Government Printing Office, September 1974), p. 5.

38. Data cited in *AgBiz Tiller*, no. 2 (September 1976), p. 3.

39. *New Republic*, October 16, 1970, p. 16.

40. Sosnick, *Hired Hands*, pp. 35–36 and the footnotes on each of these pages.

41. *Agricultural Situation*, USDA (Washington, D.C.: Government Printing Office, August 1977), p. 8.

42. Quoted in Sosnick, *Hired Hands*, p. 25. Also see Galarza, *Farm Workers and Agri-Business in California*, pp. 58–62.

43. Sosnick, *Hired Hands*, p. 43.

44. U.S., Congress, Senate, *The Migratory Farm Labor Problem in the United States*, 1969 report of the Committee on Labor and Public Welfare, Report No. 91-83, 91st Cong., 1st sess., 1969, pp. 54–55 (hereinafter *Migratory farm Labor Problem*).

45. Ibid., p. 53.

46. Sosnick, *Hired Hands*, p. 273.

47. Ibid., p. 272.

48. Ibid., p. 28.

49. Ibid., p. 29.

50. Ibid., p. 279.

51. Ibid., p. 286.

52. Ibid., p. 288.

53. *The Conditions of Farmworkers and Small Farmers* (Washington, D.C.:

National Sharecroppers Fund and Rural Advancement Fund, 1972), p. 1.

54. *Society*, April 1972, p. 40.

55. *Migratory Farm Labor Problem*, p. 34.

56. *Society*, April 1972, p. 37.

57. Coles and Davey, "Working with Migrants," pp. 14–16 and 21.

58. *Migratory Farm Labor Problem*, p. 25.

59. Ibid., pp. 31, 37, 41, and 45–47.

60. Galarza, *Farm Workers and Agri-business in California*, p. 316.

61. *The Conditions of Farmworkers and Small Farmers*, p. 3.

62. Ibid., p. 4.

63. Coles and Davey, "Working with Migrants," p. 16.

64. Joan W. Moore, "Colonialism: The Case of the Mexican Americans," *Social Problems* 17, no. 4 (Spring 1970), pp. 463–472.

65. Joel Solkoff, "Can Cesar Chavez Cope With Success?," *New Republic*, May 22, 1976, p. 14.

66. Ibid.

67. For a history of the farm labor movement, see London and Anderson, *So Shall Ye Reap*. For organized attempts to bust unionization in agriculture, see *Farmworkers in Rural America*, part 3C, pp. 1785–1792.

68. For details on the abusive powers of the labor contractor system, see *Farmworkers in Rural America*, part 3A, p. 896.

69. Peter Barnes, "Chavez Against the Wall," *New Republic*, December 7, 1974, p. 14.

70. Peter Barnes, "Cesar Chavez and the Teamsters," *New Republic*, May 19, 1973, p. 15. For a historical account of Cesar Chavez and the farm workers' movement, see Peter Matthiessen, *Sal Si Puedes* (New York: Dell Publishing Co., Delta Books, 1969).

71. Barnes, "Chavez Against the Wall," pp. 13–14.

72. "UFW Winning California Elections," *Dollars and Sense*, no. 13 (January 1976), p. 13. In February 1979 the UFWA again called for a national boycott, this time against Chiquita brand bananas. Chiquita's parent company, United Brands, owns Sun-Harvest, Inc., one of 10 lettuce companies against which the UFWA has been striking for higher wages and improved employee benefits.

73. "UFW Still Fighting Teamsters: The State of the Union," *Albatross*, April 25, 1975, p. 4.

74. "UFW Winning California Elections," p. 12. The Teamsters lost prestige, membership, and money (an estimated $7 million) in fighting a democratically worker-run union. For yet another episode in this very long and complicated struggle, see Jorge Carralejo, "Report on Proposition 14: Farmworkers vs. Big Growers, Big Money and Big Lies," *Radical America* 2, no. 2 (March–April 1977), pp. 74–78.

75. Solkoff, "Can Cesar Chavez Cope with Success?" p. 13.

76. Sam Kushner, "Farmworkers Celebrate Victory Over Teamsters, but Cannery Row Looms," *Seven Days*, April 25, 1977, p. 10. Despite the impressive accomplishments of the UFWA, less than 40,000 workers, out of a

total of more than 250,000 in California, are under contract. Several million farm workers across the country still remain to be organized (*New Republic*, July 24, 1976, letter to the editor, p. 31).

Chapter 12   The Decline of Agriculturally Dependent Small Towns

1. U.S., Congress, Senate, *Farmworkers in Rural America, 1971-1972*, *Hearings* before the Subcommittee on Migratory Labor of the Committee on Labor and Public Welfare, 92d Cong., 2d sess., 1972, part 2, p. 356.

2. Ibid., p. 357.

3. Walter R. Goldschmidt, *As You Sow: Three Studies of the Social Consequences of Agribusiness* (Montclair, N.J.: Allanheld, Osmun, & Co., 1978).

4. Walter R. Goldschmidt, "Small Business and the Community: A Study in Central Valley of California on Effects of Scale of Farm Operations" (1946), reprinted in U.S., Congress, Senate, *Role of Giant Corporations*, *Hearings* before the Subcommittee on Monopoly of the Select Committee on Small Business, 93d Cong., 1st sess., 1973, part 3A.

5. Charles P. Loomis and J. Allan Beegle, *A Strategy for Rural Change* (Cambridge, Mass.: Schenkman Publishing Co., 1975), pp. 191-195.

6. See Richard S. Kirkendall, "Social Science in the Central Valley of California," *California Historical Society Quarterly* (1964), pp. 195–218.

7. Goldschmidt, "Small Business and the Community," pp. 4571 and 4575.

8. Small Farm Viability Project, *The Family Farm in California* (November 1977), Technology Task Force report, appendix A.

9. Isao Fujimoto, "The Communities of the San Joaquin Valley," in U.S., Congress, Senate, *Priorities in Agricultural Research of the U.S. Department of Agriculture—Appendix*, Subcommittee on Administrative Practice and Procedure of the Committee on the Judiciary, 95th Cong., 2d sess., 1978, part 2, pp. 1374–1396.

10. Jan L. Flora, Ivan Brown, and Judith Lee Conboy, "Impact of Type of Agriculture on Class Structure, Social Well-Being and Inequalities" (Research paper, Kansas State University, 1977).

11. Ibid., p. 27.

Chapter 13   Farmers Challenge Agribusiness

1. Quoted in Grant McConnell, *The Decline of Agrarian Democracy* (Berkeley and Los Angeles: University of California Press, 1959), p. 5.

2. From the 1920s to the 1960s, agrarian radicalism flourished in the form of thirty-one groups and fifty-two journals. See Lowell K. Dyson, "Radical Farm Organization and Periodicals in America, 1920-1960," *Agricultural History* 45, no. 2 (April 1971), pp. 111-120.

3. S. M. Lipset, *Agrarian Socialism* (Berkeley: University of California Press, 1959), p. 5.

4. Ibid., p. 3.

5. Kirk H. Porter and Donald B. Johnson, "The Fundamental Principles of Just Government," in Louis H. Douglas (ed.), *Agrarianism in American History* (Lexington, Mass.: D. C. Heath & Co., 1969), pp. 100–102.

6. Lipset, *Agrarian Socialism*, p. 6.

7. Quoted in McConnell, *Decline of Agrarian Democracy*, p. 5.

8. Lawrence Goodwyn, *The Populist Movement* (New York: Oxford University Press, 1978), p. xvii.

9. Ibid., p. 98.

10. Ibid., pp. xviii–xxi and footnote on p. 181.

11. Ibid., p. xix.

12. John D. Hicks, "The People, Not the Plutocrats, Must Control the Government," in Louis H. Douglas (ed.), *Agrarianism in American History* (Lexington, Mass.: D. C. Heath & Co., 1969), pp. 103–111.

13. Lipset, *Agrarian Socialism*, p. 14.

14. Earle D. Ross, "Self-Help and Individual Initiative Remain . . . Typical of Life in the Country . . . ," in Louis H. Douglas (ed.), *Agrarianism in American History* (Lexington, Mass.: D. C. Heath & Co., 1969), pp. 130–137.

15. Samuel R. Berger, *Dollar Harvest* (Lexington, Mass.: Heath Lexington Books, 1971), p. 92.

16. Ibid., p. 93.

17. The Farm Holiday Association made "substantial contributions of food" to the 1934 general strike in Minneapolis led by the Teamsters. Over 35,000 building trade workers walked out in sympathy. See Jeremy Brecher, *Strike!* (Boston: South End Press, 1972), p. 162.

18. Berger, *Dollar Harvest*, p. 98. In 1936 the U.S. Supreme Court ruled that the AAA was unconstitutional and that controlling agricultural production was a local matter for the states to decide. In 1938 a new AAA was passed and the influence of the Farm Bureau continued unabated.

19. McConnell, *Decline of Agrarian Democracy*, pp. 88–89.

20. Ibid., pp. 44–65 and pp. 97–111. Also consult Berger, *Dollar Harvest*, for the most detailed examination of the bureau's conservative philosophy, authoritarian tactics, and agribusiness perspective.

21. "Family Farmers Strike for Survival," *Dollars and Sense*, no. 34 (February 1978), pp. 14–15. See also "Sizzling Statistics," *New Republic*, February 25, 1978, pp. 3 and 48.

22. Timothy Lange, "A Farmers' Strike is No Circus, It's Serious Business," *In These Times*, December 21–27, 1977, pp. 9–16.

23. "AAM Isn't Dead: Massive Tractorcade Scheduled for January," *The Country Today*, December 20, 1978, pp. 1 and 8.

24. Catherine Lerza, "Farmers on the March," *The Progressive*, June 1978, pp. 22–25.

25. Martin Brown, "Conflict of Interest Over Food," *In These Times*, January 4–10, 1978, p. 7.

26. Lerza, "Farmers on the March."

27. C. Edward Harshbarger and Marvin Duncan, "Parity—Is It the Answer?" *Economic Review*, Federal Reserve Bank of Kansas City, June 1978, pp. 3–14.

28. Richard J. Margolis, "Beneath the Harrow," *Rural America* 4, no. 3 (March 1979), p. 4.

29. Judy Strasser, "A New Breed of Farm Women Fighting to Survive," *In These Times*, December 21–27, 1977, pp. 14–15.

30. Lipset, *Agrarian Socialism*, p. 9.

31. Goodwyn, *The Populist Movement*, p. 284.

Chapter 14   Where Have All the Family Farmers Gone?
A Structural Analysis of U.S. Agriculture

1. Roger Barta, "Peasants and Political Power in Mexico: A Theoretical Approach," *Latin American Perspectives*, issue 5, vol. 2, no. 2 (Summer 1975).

2. T. G. McGee, "Peasants in the Cities: A Paradox, A Paradox, Most Ingenious Paradox," *Human Organization* 32, no. 2 (Summer 1973), p. 141.

3. T. Wharton Collens, *The Eden of Labor* (Philadelphia: Henry Baird, 1876). The following discussion is based on Collens's imagery.

4. Karl Marx and Friedrich Engels, *The Communist Manifesto* (New York: Appleton-Century-Crofts, 1955), pp. 17–18.

5. See the statement by John T. Caldwell, representing the National Association of State Universities and Land-Grant Colleges, in U.S., Congress, Senate, *Farmworkers in Rural America, 1971–1972, Hearings* before the Subcommittee on Migratory Labor of the Committee on Labor and Public Welfare, 92d Cong., 2d sess., 1972, part 4B, pp. 2760–2762.

6. Barta, "Peasants and Political Power in Mexico," p. 127.

7. Ibid., p. 131.

8. Richard Milk, "The New Agriculture in the United States: A Dissenter's View," *Land Economics* 48, no. 3 (1972), p. 230.

9. Ibid., p. 231.

10. For a discussion of landownership as monopoly power and land rent as monopoly payment, see Michael Perelman, "Natural Resources and Agriculture: Karl Marx's Economic Model," *American Journal of Agricultural Economics* 57, no. 4 (November 1975), p. 703.

11. Barta, "Peasants and Political Power in Mexico," p. 134.

12. James Iron, "How to Squeeze the Farmer Successfully," *Journal of Developing Areas* 4 (January 1970), p. 154.

13. Ibid., p. 155.

14. Karl Marx, *The Eighteenth Brumaire of Louis Bonaparte* (New York: International Publishers, 1963), p. 127.

15. Jeffery M. Paige, *Agrarian Revolution* (New York: Free Press, 1975), p. 13.

16. Radoje Nikolitch, *Family-Size Farms in U.S. Agriculture*, USDA, Economic Research Service Report No. 499 (Washington, D.C.: Government Printing Office, 1972), p. 34.

17. Harry Cleaver, *Reading Capital Politically* (Austin: University of Texas Press, 1979), p. 165. For a detailed discussion of women's contribution toward the maintenance and reproduction of labor power, see Carmen Diana

Printed and bound by CPI Group (UK) Ltd, Croydon, CR0 4YY
6178/2023/3/1010290

Deere, "Rural Women's Subsistence Production in the Capitalist Periphery," *Review of Radical Political Economics* 8, no. 1 (Spring 1976), pp. 9–17.

18. Cleaver, *Reading Capital Politically*, p. 109.

19. See ibid., p. 74, for Karl Marx's distinction between a class-in-itself and a class-for-itself.

20. The following national labor unions and/or locals and affiliates are supportive of democratic socialism: Buildings and Construction Trade Department, AFL-CIO; United Automobile Workers; United Steelworkers of America; National Union of Hospital and Health Care Employees; Amalgamated Clothing and Textile Workers Union; United Automobile, Aerospace, and Agricultural Implement Workers Union of America, UAW; International Association of Machinists and Aerospace Workers; Coalition of Black Trade Unionists; and International Chemical Workers Union.

21. One such progressive farmer is Roger McAfee, a thirty-three-year-old white farmer, who put up 405 acres of his family's 1,100-acre cooperative farm near Fresno, California, as collateral for most of Angela Davis's bail bond. "I'm just a working man who agrees with the philosophy of Angela Davis," he said (*Time*, March 6, 1972, p. 26).

22. Michael Harrington, "Getting Restless Again," *New Republic*, September 1 and 8, 1979, pp. 12–15.

23. For its positions on energy, inflation, labor, human rights, health, education, family farming, international justice, and democratic socialism, contact the Farmer-Labor Association, 3200 Chicago Avenue, Minneapolis, MN 55407.

Index